Gas Turbines

Gas Turbines

Editors
_____

Suhas Joshi and Bhushan Kulkarni

Gas Turbines

Edited by **Suhas Joshi and Bhushan Kulkarni**

Printed in 2017

ISBN: 978-1-68117-494-5

Library of Congress Control Number: 2015936534

© 2016 by
SCITUS Academics LLC,
616, Corporate Way, Suite 2, 4766,
Valley Cottage, NY 10989

www.scitusacademics.com

Contents

Preface vii

Chapter 1 Gas Turbines: Gas Cleaning Requirements for Biomass-Fired
Systems 1
John Oakey; Nigel Simms; Paul Kilgallon

Chapter 2 Technology Review of Modern Gas Turbine Inlet Filtration
Systems 23
Melissa Wilcox, Rainer Kurz, and Klaus Brun

Chapter 3 Control System Design for a Gas Turbine Engine Using
Evolutionary Computing for Multidisciplinary Optimization...........63
Valceres V. R. e Silval, Wael Khatibll, and Peter J. Flemingll

Chapter 4 The Thermal Stability of Eyjafjallajökull Ash versus Turbine
Ingestion Test Sands 81
Ulrich Kueppers, Corrado Cimarelli, Kai-Uwe Hess,
Jacopo Taddeucci, Fabian B Wadsworth, and Donald B Dingwell

Chapter 5 Optimum Parametric Performance Characterization of an
Irreversible Gas Turbine Brayton Cycle 107
Maher M Abou Al-Sood, Kassem K Matrawy, and
Yousef M Abdel-Rahim

Chapter 6 Development of Semiclosed Cycle Gas Turbine for Oxy-Fuel
IGCC Power Generation with CO_2 Capture 135
Takeharu Hasegawa

Chapter 7 Gas Turbine Cogeneration Groups Flexibility to Classical and
Alternative Gaseous Fuels Combustion 169
Ene Barbu, Romulus Petcu, Valeriu Vilag, Valentin Silivestru, Tudor
Prisecaru, Jeni Popescu, Cleopatra Cuciumita, and Sorin Tomescu

Chapter 8   **Review of the New Combustion Technologies in Modern Gas
                  Turbines** **207**

                  M. Khosravy el_Hossaini

                  Citations..**233**
                  Index..**237**

Preface

A gas turbine, also called a combustion turbine, is a type of internal combustion engine. It has an upstream rotating compressor coupled to a downstream turbine, and a combustion chamber in-between. The basic operation of the gas turbine is similar to that of the steam power plant except that air is used instead of water. Fresh atmospheric air flows through a compressor that brings it to higher pressure. Energy is then added by spraying fuel into the air and igniting it so the combustion generates a high-temperature flow. This high-temperature high-pressure gas enters a turbine, where it expands down to the exhaust pressure, producing a shaft work output in the process. The turbine shaft work is used to drive the compressor and other devices such as an electric generator that may be coupled to the shaft. The energy that is not used for shaft work comes out in the exhaust gases, so these have either a high temperature or a high velocity.

Editor

Chapter 1

Gas Turbines: Gas Cleaning Requirements for Biomass-Fired Systems

John Oakey; Nigel Simms; Paul Kilgallon

Power Generation Technology Centre, Cranfield University, Bedford, UK

ABSTRACT

Increased interest in the development of renewable energy technologies has been hencouraged by the introduction of legislative measures in Europe to reduce CO_2 emissions from power generation in response to the potential threat of global warming. Of these technologies, biomass-firing represents a high priority because of the modest risk involved and the availability of waste biomass in many countries. Options based on farmed biomass are also under development.

This paper reviews the challenges facing these technologies if they are to be cost competitive while delivering the supposed environmental benefits. In particular, it focuses on the use of biomass in gasification-based systems using gas turbines to deliver increased efficiencies. Results from recent studies in a European programme are presented. For these technologies to be successful, an optimal balance has to be achieved between the high cost of cleaning fuel gases, the reliability of the gas turbine and the fuel flexibility of the overall system. Such optimisation is necessary on a case-by-case basis, as local considerations can play a significant part.

INTRODUCTION

Increased interest in the development of renewable energy technologies has been encouraged by introduction of legislative measures in Europe to reduce CO_2 emissions from power generation in response to the potential threat of global warming. Of these technologies, biomass-firing represents a high priority because of the modest technological risk involved and the availability of waste biomass in many countries[1]. Options based on farmed biomass are also under development.

While combustion of waste and farmed biomass has been practised for many years around the world, system efficiencies have always fallen well below those of equivalent fossil-fired systems. In most cases this has been due to the reduced steam conditions enforced by the severe fouling and corrosion problems experienced as a result of the high contaminant levels (e.g. Na, K, Cl, Pb, etc) with many biomass fuels which also lead to reduced component lives. While coal plants are targeting 650 °C/300bar steam and above, biomass plants are currently operating at less than 540 °C/100 bar steam with efficiencies of typically less than 30%.

In Denmark, where government legislation has driven the introduction of ever more efficient plants, the most advanced straw-fired biomass plant operate at 540 °C/92 bar with an electrical efficiency of 29%. Experience from boilers in Sweden firing 100% forest fuel, indicates that conventional superheater steels last no longer than four years or 20,000 h before they must be replaced because of corrosion damage. Overall, this leads to higher operating costs making biomass combustion plants uncompetitive compared to fossil plants, unless supported in some way through subsidies or grants. Moving towards cheaper waste biomass sources (such as demolition wood) to improve the overall plant economics, has been found to lead to even more severe problems. In addition to the impact of biomass fuels on operating costs, the capital costs of biomass plants are usually higher than their fossil counterparts due to more complex fuel feeding arrangements, fuel drying, gas cleaning, emissions monitoring requirements, etc.

In order to improve system efficiencies and improve the economics of biomass plants, recent interest has focused on gasification combined cycle systems which use turbines and advanced gas engines. While different gasification options are possible[2], circulating fluidised bed gasifiers have been developed to the greatest extent due to their flexibility and suitability for the scale of available biomass feedstocks. Both atmospheric pressure and pressurised schemes have been demonstrated at a scale using a small industrial gas turbine (e.g. an ALSTOM Power ~4MW$_e$ Typhoon).

Based on the TPS Termiska Processer AB circulating bed gasification system[3] from Sweden, the world's first 'commercial' biomass gasification plant (known as the ARBRE project) is undergoing commissioning at Eggborough, Yorkshire in the UK. This plant uses coppiced willow and forestry residues in chipped form and produces 8MW of electricity with a cycle efficiency of ~31%. A generic flowsheet for the hot gas path of this plant through to the gas turbine is shown in Fig. 1.

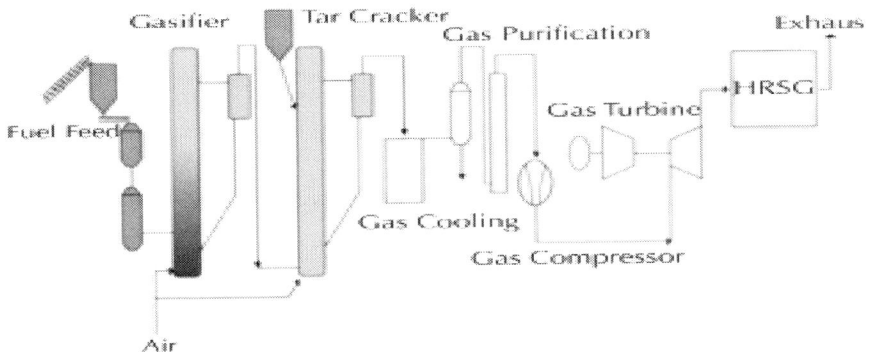

Figure 1: Simplified flowsheet for the hot gas path of the ARBRE project.

This figure shows the complexity of the hot gas path in such systems. Even though fluidised bed gasifiers lead to moderate fuel gas tar levels, a high temperature cracker is used to reduce energy losses and to limit the tar removal burden at the gas purification/scrubbing stage. Tars would otherwise cause problems in the gas compressor. Ammonia levels in the fuel gas are a further concern as they will lead to excessive NO_x levels in the gas turbine exhaust and exceed the allowable emissions limits.

In a pressurised system, such as the that in the Varnamo project[4] operated by Sydkraft AB using Foster Wheeler gasifier technology from Finland, there is no requirement for a fuel gas compressor (see Figure 2). So, tars can be kept hot (in the vapour phase) provided they do not exceed the gas turbine entry limits and do not cause blinding problems in the hot gas filter. Being at high pressure and using a hot gas cleaning approach reduces the complexity of the hot gas path and raises the cycle efficiency. The measured efficiency in the Varnamo project was 32% but up to 38% could be expected from application of the latest gas and steam turbine technology. This scheme used a variety of biomass fuels to demonstrate its flexibility and produced 5 MW of electricity and 9 MW of heat for district heating. Figure 2 shows the hot gas path through to the gas turbine.

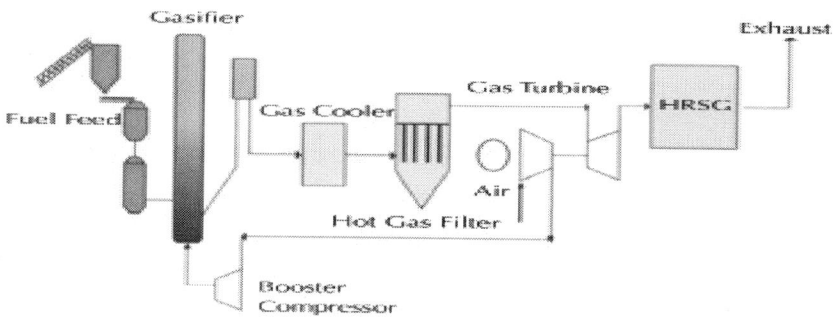

Figure 2: Flowsheet for the hot gas path of the Varnamo project.

So, while the gasification approach leads to higher efficiencies, it is more complex and expensive to build. It is also susceptible to problems associated with the same contaminants which have led to the operational restrictions experienced with biomass combustion plant. The remainder of this paper reviews the possible effects of contaminants in biomass gasification systems as described above, with particular reference to the durability of the gas turbine and the implications this may have for gas cleaning requirements.

BIOMASS CHARACTERISTICS AND EFFECTS ON FUEL GAS CONTAMINANTS

Like coal, biomass contains a wide range of elements that may react to form potentially harmful deposits in gasification systems and their gas turbines. The 'mix' of elements in the fuel gases produced in a gasification process will be highly dependent on the biomass fuel composition. Before biomass-firing can be used with any confidence in gasification systems, it is necessary to investigate the effects the deposits and gas environments will have on the gas turbine components in such systems. From such information, fuel specifications for biomass-fired gas turbines can be derived, to ensure adequate lives for components and to permit the use of state-of-the-art gas turbines.

In order to investigate the contaminant effects in the hot gas path of biomass gasification plants, it is necessary to understand the levels in biomass fuels relative to those in coals for which there is wide experience in combined cycle gasification systems. Extensive composition information has been gathered on potential European biomass fuels[5]; but few of these analyses have been carried out for all minor and trace metal species. Average data values for pine wood, wheat straw, a range of grasses, sewage sludge and peat, with coal for comparison are given in Tables 1 and 2; but it should be noted that there are significant differences in the errors associated with each of these values due to inherent fuel variations, the varying numbers of references used to determine each value and the various analytical methods used.

Table 1: Average analyses of biomass and fossil fuels

	Wood	Wheat Straw	Grass	Sewage Sludge	Coal
Moisture (wt%)	20.7	101	14.9	19.5	8.2
Ash (wt%)	1.7	5.9	5.2	43.4	12.7
S (wt%)	0.2	0.1	0.2	1.0	1.7
Cl (wt%)	0.1	0.8	0.2	0.1	0.2
LHV, MJ/kg	18.6	17.3	18.3	10.7	26.2

Table 2: Ash analyses of biomass and fossil fuels

	Al_2O_3	SiO_2	Na_2O	K_2O	MgO	CaO	Fe_2O_3	P_2O_5	SO_3	TiO_2
Wood	5.5	24.3	1.7	9.3	4.5	34.5	3.6	5.6	5.5	0.4
Wheat Straw	1.8	49.6	3.7	22.2	2.9	6.0	1.0	2.6	3.3	0.1
Grass	2.8	59.5	0.7	15.3	3.4	7.4	1.6	8.6	1.4	0.2
Sewage Sludge	15.0	34.6	1.0	1.4	3.1	17.3	10.6	10.0	1.3	1.0
Coal	18.1	40.8	3.5	2.4	3.8	10.3	12.3	6.2	6.2	0.8

Table 1 illustrates the major differences between biomass fuels and coal. In general terms, they have higher moisture, lower ash (except sewage sludge), lower S and similar or higher Cl. Table 2 presents average ash compositions derived using standard ash analysis techniques derived to give comparable data for combustion systems. Noting the differences in ash contents from Table 1, it is also generally true that biomass fuels contain higher levels of alkali metals, in particular K. While, it must be noted that the artificial method used to generate ash for analysis was designed for coal combustion, the general findings listed do suggest that fouling and corrosion problems should be expected in biomass systems. Overall, there is a tendency for the higher Cl/lower S levels to favour the formation of chlorides over sulphates while the lower ash contents provide less dilution of any deposits formed on plant components.

It is thought that the high K content combined with Cl is also responsible for the formation of low melting temperature compounds during combustion; the low S content in many biomass fuels is another contributing factor. These low melting point ash constituents have led to the widespread fouling and severe corrosion problems experienced.

In gasification systems, the situation is somewhat different. There are several possible routes for minor and trace elements, such as Na or K, to take within a gasification system and through into the gas turbine. These routes vary from no response to the gasification process (and so exit with the ash/char/slag) through to the formation of vapour species that can pass through the whole hot gas path (and so be emitted from the process). In between these two extremes, it is possible for reactions to take place forming (a) condensed particles and (b) vapour species that can condense onto entrained particles or plant components (depending on their specific operating conditions) along the hot gas path. The fate of the various trace elements is element-specific and in addition can be influenced by both the relative and absolute levels of other elements present in the fuels (e.g. S and Cl) and in any sorbents or catalysts used, as well as the composition of materials used for hot gas path components.

In order to determine the potential fate of trace elements within gasification systems using biomass fuels, one approach is to investigate the thermodynamic equilibrium at various process stages. The thermodynamic analysis package, MTDATA, has been used to predict trace element behaviour followed by comparison of the trends identified with reported plant data and known operating experience. One of the aims of this work was to enable studies to be more closely focused on realistic deposit compositions when carrying out corrosion testing on materials intended for use within gasifier and gas turbine hot gas paths.

The thermodynamic study was carried out to determine which trace elements were more/less likely to enter the hot gas paths of gasification systems, condense onto system components and/or pass through into the gas turbine. This study investigated the stability of potential product compounds in gasifier fuel gases and their sensitivity to a number of important process variables:

- two example gasifier processes: an oxygen blown entrained flow process and an air blown fluidised bed process[2,6]
- atmospheric and pressurised operation
- temperature ranges covering gasification and hot gas cleaning processes, as well as component operating temperatures
- a range of S and Cl levels to cover the potential ranges of fuels in coal and coal/biomass fired systems
- the elements As, B, Ba, Be, Ca, Cd, Co, Cu, Hg, K, Mn, Mo, Na, Pb, Sb, Se, Sn, V, Zn (Cr, Ni and Fe were not investigated as they are major alloying elements in materials used in components throughout the fuel gas paths).

Published literature surveys[7-10] were critically evaluated and care taken to avoid the pitfalls identified. For a power plant, it is important to note that kinetic effects may arise due to short gas residence times and/or slow reaction rates that could limit movement towards thermodynamic equilibrium. These effects apply especially to the bulk gases and will be less significant in the slower moving boundary layers adjacent to components.

The results of this thermodynamic study are reported elsewhere[11]. Table 3 lists the major gaseous and condensed phases and the temperature ranges for transitions between the gaseous and condensed states in gasification conditions. It should be noted that even elements with condensed phases can have significant vapour pressures. Table 4 groups the elements in terms of their 'volatility', i.e. in order of the transitions from gaseous to condensed phases. This table does not correspond with the frequently quoted three-group classification of trace elements[10]. This classification was originally developed for combustion systems and some reports directly translate this to gasification systems. As sometimes noted before[7,9] and found in this study, the same classification of elements is not applicable to combustion and gasification systems (and indeed there are several significant differences between types of gasification system).

Table 3: Summary of trace and alkali metal behaviour in gasifier gases

Element	Major Gas Species	Major Solid Species	Gas —> Solid Transformation Temperature Range (°C)
As	As, As_2, As_4, AsS		
As (+Ni)	AsS (As)	As_2Ni_5, As_8Ni_{ii}	1020-1460
B	BHO_2, $B(OH)_3$		
B (+Ca)	$B(OH)_3(BHO_2)$	$B_2Ca_3O_6$	Solid®gas 420-840
Ba	$BaCl_2$, BaC1HO	$BaCl_2$, BaS $BaCO_3$	900-1040 (or >1200)
Be	$^{Be}H2^O2$	BeO	840-960
Ca	$CaC1_2$, CaS, $CaCO_3$		
Cd	Cd, $CdCl_2$	CdS	400-540 (or <400)
Co	$CoC1_2$ (Co)	Co, Co_9S_3	600-1340
Cu	CuCl, Cu_3Cl_3 (Cu,CuH)	Cu, Cu_2S	520-960
Hg	Hg		
K	KCl, K_2C1_2	KCl	700-900
Mn	$MnC1_2$ (MnCl)	MnO, MnS	440-1260
Mo	$MoC10_2$, $MoCl_2O$, $MoC1_2O_2$ ($MoHO_2$)	MoS_2	700-1200
Na	NaCl, Na_2Cl_2	NaCl	670-900

Pb	Pb, PbCl, PbCl$_2$PbS	Pb, PbS	560-640 (or <400)
Sb	SbC1(Sb)		
Se	SeH$_2$(SeH)		
Sn	SnS, SnCl$_2$	SnO$_2$, SnS	460-560 (or <400)
V	(VCl$_2$, VCl$_3$, VOC1$_3$)	V$_2$O$_3$	
Zn	Zn, ZnCl,	ZnS	460-780

Table 4: Predicted volatility of trace and alkali metals in gasification gases

Increasing Volatility	Element
⬆	Hg, Sb, Se (As, V, B)
	Cd, Pb, Sn, Zn (As, B)
	Co, Cu, K, Mn, Mo, Na
	As, Ba, Be
	Ca (V, As, B)

Figure 3 provides an example of the data generated during the thermochemical modelling; it shows the effect of increasing the pressure on the K equilibrium diagram for an air-generated fuel gas. The gas to solid transformation temperature of the major species, KCl, is increased from ~700 °C to ~900 °C when the pressure increases from 1 to 20 atm. Thus, subject to kinetic effects, increasing pressure can be expected to lead to condensation of KCl on higher temperature surfaces in the gasifier hot gas path, reducing the remaining levels of K vapour species in the fuel gas.

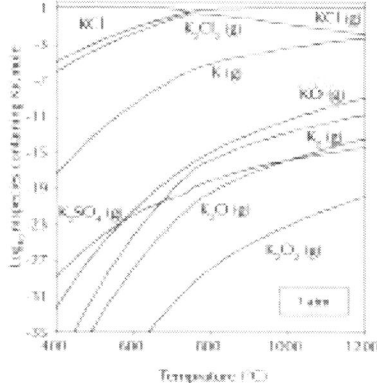 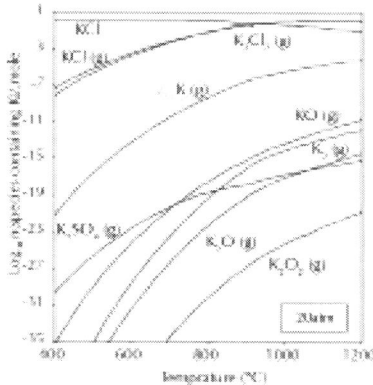

Figure 3: The effect of pressure on K equilibrium diagrams in air-derived fuel gases.

Other gas species can also influence condensation (dewpoint) temperatures. For example, increased levels of H_2S can give, for different elements, higher or lower dewpoints (e.g. Zn or Sn) or different gas phase and/or condensed species (e.g. Pb), so care is needed to ensure that the fuel gas compositions being modelled are complete and realistic.

To substantiate the effects identified by thermodynamic modelling, relevant process data were sought. There are some data for the removal of alkali, trace metals and chlorine species in gasification systems incorporating hot gas filters[7,12-17]. These data are not comprehensive and as each gasification system has used different filtering temperatures, it is difficult to separate process differences from differences arising from the use of different filter temperatures. However, it is reasonable to assume that reducing filter operating temperatures will reduce the passage of vapour phase contaminants through to the gas turbine, as it will drive the equilibrium towards condensed phases and reduce the vapour pressure of the species remaining in the gas.

In many cases (e.g. for alkali metals, Zn, etc.), the preferred route of removal on economic grounds is by condensation/reaction with the fine particles present in the gas stream followed by particle removal by the hot gas filter. However, the coldest component in this hot gas path will be the heat exchanger that cools the gases prior to entry into the filter. The presence of 'trace' metals in heat exchanger deposits at levels of several wt% has been frequently observed during component examinations and assessments (but only occasionally reported[18,19]). The large internal surfaces of the heat exchanger, ductwork and the filter unit itself provide potential sinks for condensed alkali/trace metals, especially if they are cooler than the gas stream. In a pressurised air blown fluidised bed gasifier pilot plant[7], deposits of the more volatile 'trace' metals at levels of several wt% were observed on the surfaces of pipes on the 'clean' side of the filter unit. In addition, a reduction in S and Cl-containing species has been noted across hot gas filter units, presumably as a result of reactions with the filter cake formed on the dirty side of the filter units.

The information available for alkali and trace metal species all show that lowering the filter temperature reduces the amount of vapour phase species present in the remaining fuel gas stream (and enriches the filter fines in these elements) [7,20]. Unfortunately, filter operating temperatures cannot just be lowered to the levels needed to remove most of these vapour species, as the lower limits for filter operation are dictated by other gasification processes parameters: e.g. tars and ammonia-derived compounds need to be kept in the vapour phase to avoid blinding of the filters. The available data are most comprehensive for alkali species on which many studies have been targeted. This is despite the importance of other trace elements, e.g. Pb, which are present at similar, if not higher, levels in gasification fuel gas streams and are also important in determining gas turbine component lives[7,14-16]. In work carried out at VTT[13], the most significant levels of elements found in the fuel gas were for Cd, Pb and Zn. The levels of several trace metals (e.g. Pb) and alkalis (e.g. Na, K) reported in these studies are higher than those currently acceptable for gas turbines, and in some cases it is likely that the true values of vapour phases species in the gasifier product gases are higher still[7].

From plant data, filter temperatures below 400 °C may be required to reduce contaminants to acceptable levels for the gas turbine. In other cases, supplementary gas cleaning stages or the use of a scrubber may be required or the use of a scrubber to meet the required target, impacting significantly on costs and cycle efficiency. But the level acceptable to the turbine depends on the life required of gas turbine parts, the temperatures at which they work and the materials/coatings used, so there is potential scope for compromise here also. This is discussed later in this paper.

The main points arising from the above analysis are summarised as follows: (i) fuel composition: on a mass comparison basis, the critical contaminants species (S, Cl, alkalis and trace metals) of many biomasses are similar or lower than coal - notable exceptions are K and Cl in straw; P, Cd, Pb and Zn in wood (as well as Ba, Cr, Cu, Mn); V and Zn in sewage sludge (as well as Cr, Cu, Mn, Hg); (ii) fuel gas compositions vary significantly between gasification systems at the trace contaminant level (as well as the frequently reported bulk gas composition level); (iii) most trace and alkali metals are more volatile in gasification systems than in combustion systems - the same classification of volatility as for combustion gases is not applicable (different species are volatile); (iv) Sand Cl levels (both absolute and relative), as well as operating pressure and gasification process can influence the volatility of trace and alkali metal species; (v) potentially damaging levels of Pb, Zn, Cd and Sn (and V in some systems) can all pass through the fuel gas path to the gas turbine, as well as alkali metals (Hg, B, Sb and Se can also pass through the gas turbine); and, (vi) the high levels of trace metals present in gasifier product gases can be reduced by use of low filter operating temperatures (e.g. 250 - 450 °C dependent on the gasification system).

CONTAMINANT EFFECTS ON GAS TURBINE COMPONENTS

In all biomass combined cycle systems the performance of the gas turbine is vital to the overall plant efficiency and economic viability. However, within these systems hot corrosion and/or erosion are likely to be life limiting for the gas turbine vanes and blades, rather than the creep and fatigue processes that limit their lives in the longer term due to the expected contaminant levels.

The environments found within the hot gas paths of gas turbines depend on the contaminants present in the fuel and air entering the turbine, as well as the turbine operating conditions. Industrial gas turbines have been developed to fire on a wide variety of fuels, ranging from natural gas to sour gases and heavy fuel oils. The degradation of materials in such systems has been the subject of many investigations during the past 40 years, as operating conditions have developed and/or fuels have changed, and the potential problems which many be encountered in gas and oil fired gas turbines have been well characterised[21,22].

Many similar types of materials degradation can be expected in gas turbines using solid fuel derived fuel gases, as some of the contaminant species are the same as for oil and/or gas fired systems. However, the contaminant levels are different, there are additional as well as absent species and the sources/forms of the contaminant species also differ. Fuel gases derived from biomass have the potential to cause both erosion and corrosion damage to gas turbine hot gas path components. Fuel derived particles can cause either erosion damage or deposition depending on the particles' size and composition, as well as aerofoil design and operating conditions. Corrosion can result from the combined effects of gaseous species (e.g. SO_x and HCl) and deposits formed by condensation from the vapour phase (e.g. alkalis and other trace metal species) and/or particle impaction and sticking.

The mode of corrosion damage is highly dependent on the local component environment. Conventionally, the metal vapour species of most concern were alkalis (mainly Na) and S in gas turbines fired on clean fuels (either as fuel contaminants or via the combustion air) or V from heavy fuel oils. In biomass fired systems the levels of both SO_x and HCl can be similar to (or higher for HCL) than those from a coal gasifier (see Table 1). Also, the fuel gas may contain significant levels of alkali metals (in particular K) and heavy metals, e.g. Pb and Zn[23], depending on the type of biomass (Table 2) and the effectiveness of the gas cleaning approach. If a water or chemical scrubber is included (as in the ARBRE scheme for ammonia removal), the levels of these contaminants will be significantly reduced whereas a hot dry cleaning approach will potentially lead to higher levels, depending on the operating temperature of the filter.

The effects of different fuels on the levels of contaminant vapour phase species, deposition fluxes and deposit compositions (e.g. melting points) requires careful consideration for each process and fuel. Thus, fuel and air quality standards produced for gas turbines fired on more traditional fuels [e.g. 24] need to be thoroughly reviewed and revised[25-26], to take into account the significant differences with these new fuel compositions, as well as the damage rates that will be acceptable for these new power systems.

In order to identify whether the residual levels of contaminants (such as K) in a 'cleaned', biomass-derived fuel gas will restrict the operating life of gas turbine components, it is necessary to understand their deposition behaviour and the effects the resultant deposit may have on the blade and vane materials and coatings.

Thermodynamic analysis methods were also applied to the gas turbine, using the results of the gasifier fuel gas study to identify the elements that can pass through into the gas turbine, as well as realistic ranges for the different contaminant levels. Figure 4 illustrates the results obtained for K dewpoint temperatures, which are plotted as a function of contaminant levels for different SO_x and HCl levels and gas pressures. In a gas turbine combustion gas, K_2SO_4 is the equilibrium K compound in deposits rather than KCl which will persist in the gas phase. Figure 4 shows that the K_2SO_4 dewpoint increases with increasing K or SO_x levels in the gas stream, as well as gas pressure, but decreases with increasing HCl levels in the gas stream.

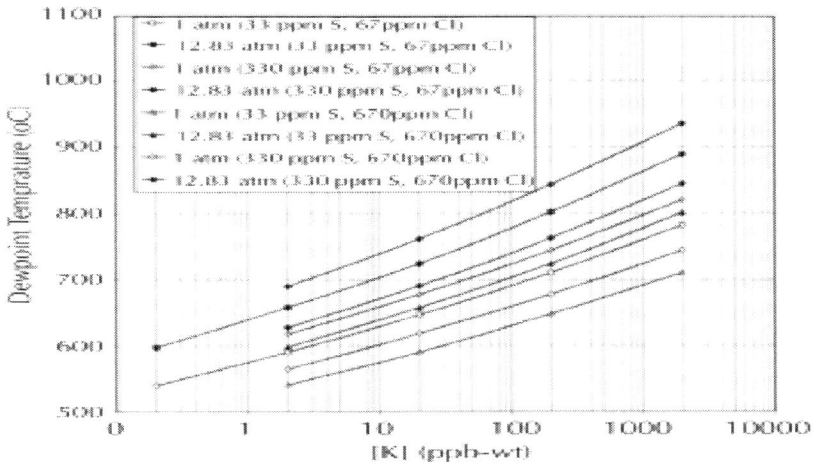

Figure 4: Dependence of K_2SO_4 dewpoint on gas pressure, S, Cl and K levels.

Once deposited, there are significant differences between the effects of the various deposits; described in detail elsewhere[27,28]. This is illustrated in Fig. 5 for a 'deposition flux' of 5 µg/cm²/h, in which 'maximum' corrosion damage values (damage with a 4% probability of being exceeded) are plotted as a function of exposure time(data from laboratory corrosion tests). The samples with this 'deposition flux' at 700 °C have a clear incubation period for all the various deposit compositions. However, this incubation period is significantly shorter for the Pb/Zn containing deposits, than the alkali sulphate deposits, though there are still significant differences in damage with varying Na/K ratios in these ‹deposits› after 2000 h testing. Similar deposit composition effects were observed at other deposition fluxes[27].

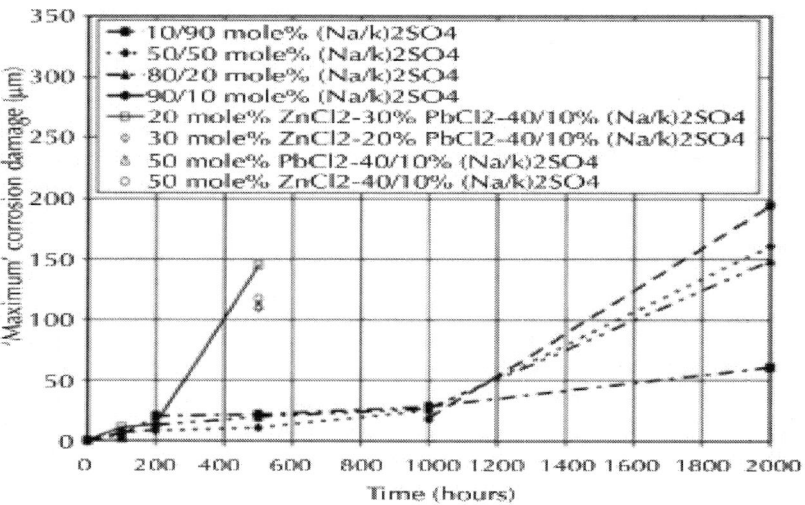

Figure 5: Effect of different deposit compositions on 'maximum' corrosion damage to IN738LC at 700 °C (deposition flux = 5 µg/cm² /h, SO$_x$ = 2000vpm, HCl = 350 vpm)[29].

The effect of deposition flux on the maximum corrosion damage obtained for IN738LC at 700 °C with a deposit mix of 80/20 mole% (Na/K)$_2$SO$_4$ is illustrated in Fig. 6. This is typical of the response of gas turbine materials, with the dependence of corrosion rate on deposition flux being approximately sigmoidal with three distinct behaviour regimes[30]:

-at low deposition fluxes (< ~1 µg/cm²/h) there were low corrosion rates;

-at intermediate deposition fluxes (from ~1 to ~30 µg/cm²/h), much higher corrosion rates were found with a dependence on deposition flux close to linear;

-at high deposition fluxes (> ~30 µg/cm²/h), a thick scale/deposit layer was found which stabilises or even slightly reduces the corrosion rate with further increases in flux.

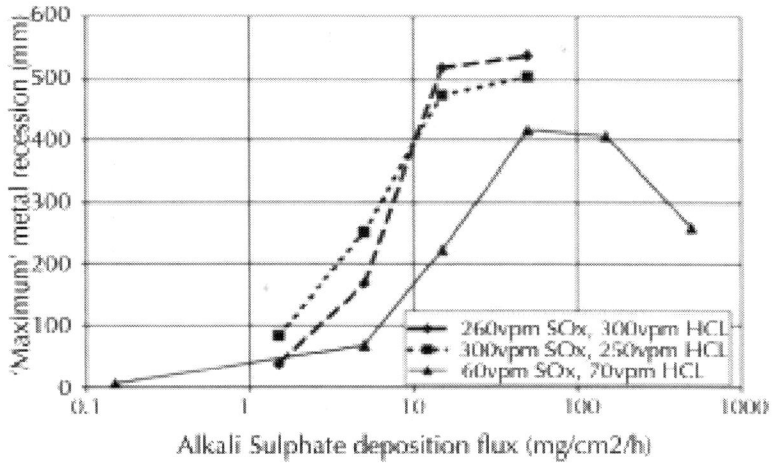

Figure 6: Effect of alkali sulphate flux and gaseous reactants on 'maximum' metal damage recession of IN738LC at 700 °C[29].

The effect of varying SO_x and HCl concentrations on the hot corrosion damage of IN738LC at 650 and 700 °C were evaluated from a series of tests in which each of these contaminants was varied for the same ‹deposit› composition (80/20 mole % $(Na/K)_2SO_4$) and a series of 'deposition fluxes'. It was found that the effect of SO_x levels was much more significant than HCl[27,30].

From the above, it is clear that great care is required in the control of trace alkali and other contaminants because of their potential to cause excessive corrosion damage to the hot gas path components of a gas turbine or gas engine using a biomass-derived fuel gas. The exact level of damage that can be tolerated depends on specific process and operating parameters (e.g. temperatures, materials, repair/replacement strategy etc.), as well as economic factors. It may be possible to optimise the materials selection and operating conditions of a gas turbine, in combination with a comprehensive maintenance strategy (to allow it to handle higher than currently specified contaminant levels) instead of bearing the high cost of an elaborate gas cleaning system.

CONCLUSIONS FOR BIOMASS COMBINED CYCLE SYSTEMS

From the above, it is clear that care is needed in designing the process flowsheets for biomass systems. In particular, those for fuels such as wheat straw, with very high K and Cl levels, require careful consideration of the removal of fuel contaminants so that a reliable and cost effective gas cleaning approach can be adopted.

For the lower efficiency atmospheric pressure systems, scrubbing to remove some of the residual tar and ammonia should also reduce significantly the water soluble trace contaminants such as KCl. But as the gas is cooled substantially prior to scrubbing to minimise energy losses and to meet scrubber entry requirements, many of the trace contaminants may well have condensed onto entrained particulates or component surfaces, before the gas reaches the scrubber. To minimise costs, these systems do not have filters prior to the scrubber as for most large coal-fired IGCC plants, so the dissolved contaminants and the particulates will end up in the scrubber discharge. From the gas turbine perspective, the scrubbed fuel gas should not present any problems as nearly all fuel-borne contaminants should have been removed.

For the higher efficiency, pressurised systems appropriate hot, dry gas cleaning schemes are required to avoid the higher costs of scrubbing the high pressure fuel gas. Filtration is the key element, allowing the combination of particulate removal with condensed trace contaminants dependent on the filtration temperature. Filtration also allows for the use of in-duct injection of sorbents to reduce alkalis and HCl with minimal extra complication. However, because of the need to retain any uncracked residual tars in the vapour phase and the need to avoid condensation of ammonium compounds (both of which would lead to blinding of the filter elements), filtration temperatures cannot be reduced excessively in the drive to limit trace species. In these schemes, costly additional catalytic NO_x reduction measures either upstream or downstream of the gas turbine may be needed. The gas turbine in these systems will be passing reduced levels of fuel-borne contaminants which pass the gas cleaning stages. There is scope to adjust the operation and materials of the turbine to suit the biomass fuel used; while, use of the highest efficiency turbine is an attractive option, it may prove uneconomic when the required gas cleaning costs are considered.

REFERENCES

1. Energy for the Future: Renewable Sources of Energy - Campaign for Take-off - a Community Strategy and Action Plan, EU Commission Paper, 1998

2. Bridgwater, A.V. Biomass Gasification for Power Generation', *Fuel*, v. 74, n. 5, p. 631-653, 1995

3. Rensfelt, E. Gasification Technology for Biomass and Other Solid Fuels, *Biosolids Energy Recovery Technology Seminar*, IWEX 2001, Birmingham UK, 2001

4. Stahl, K.; Neergaard, M. IGCC Power Plant for Biomass Utilisation, Varnamo, Sweden, *Biomass and Bioenergy*, v. 15, n. 3, 1998

5. 'Gas turbines in advanced co-fired energy systems', ECSC Project 7220-PR/053

6. Takematsu, T.; Maude, C.W. Coal Gasification for IGCC Power Generation, *IEACR/37*, IEA Coal Research, London, UK (1991).

7. Reed, G.P. *Control of Trace Elements in Gasification*, Ph.D. Thesis, Imperial College, London (2000).

8. Frandsen, F.; Dam-Johansen, K.; Rasmassen, P. Trace Elements from Combustion and Gasification of Coal an Equilibrium Approach, *Prog. Energy Combustion Sci.*, v. 20, p. 115-138, 1994

9. Fantom, I.R. An Assessment of Trace Element Concentrations in the British Coal Topping Cycle, *Power Generation Report,* n. 121, CRE, British Coal, 1991

10. Clarke, L.B.; Sloss, L.L. Trace elements - emissions from coal combustion and gasification, *IEACR/49*, IEA Coal Research, London, 1992

11. Kilgallon, P.J.; Simms, N.J.; Oakey, J.E. Fate of Trace Contaminants from Biomass Fuels in Gasification Systems, to be published in *Materials for Power Engineering*, Liege, 2002

12. 4th International Conference on Gas Cleaning at High Temperatures, Karlsruhe, Germany (Sept. 1999).

13. Nieminen, M. *et al.*, Raskasmetallien käyttäytyminen turpeen ja kivihiilen leijukerros- kaasutuksessa,*LIEKKI-Combustion Research Program*. Project Report, 1991

14. Kurkela, E. *et al.* Pressurized fluidized-bed gasification experiments with biomass, peat and coal at VTT in 1991-1994. Part 3. Gasification of Danish wheat straw and coal, *VTT Publication*, v. 291, 1996

15. Kurkela, E. *et al.* Pressurized fluidized-bed gasification experiments with wood, peat and coal at VTT in 1991-1992. Part 1. Test facilities and gasification experiments with sawdust, *VTT Publication*, v. 161, 1993

16. E. Kurkela *et al.*, 'Pressurized fluidized-bed gasification experiments with wood, peat and coal at VTT in 1991-1994. Part 2. Experiences from peat and coal gasification and hot gas filtration'. VTT Publication 249, (1995).

17. 'Integrated hot fuel gas cleaning for advanced gasification combined cycle processes'. LIEKKI- Combustion and Gasification Research Programme. LIEKKI- vuosikirjat, projekti: 410 (1996).

18. Proc. Corrosion in Advanced Power Plants, Special Issue of Materials at High Temperatures, vol. 14(1997).]

19. Proc. First International Workshop on Materials for Coal Gasification Power Plant, Special Issue of Materials at High Temperature, vol. 11 (1993).

20. Mitchell, S.C. Hot gas clean-up of sulphur, nitrogen, minor and trace elements, *IEA Coal Research*, UK, 1998.

21. Sims, C.T.; Stoloff, N.S.; Hagel, W.C. *Superalloys II*, Wiley (1987)

22. 'Hot Corrosion Standards, Test Procedures and Performance', High Temperature Technology vol. 7 (4)(1989).

23. 'Co-gasification of Coal/Biomass and Coal/Waste Mixtures', Final Report EC APAS Contract COAL-CT92-0001, University of Stuttgart, Germany (1995).

24. ASTM D2880, Standard Specification for Gas Turbine Fuel Oils (1990).

25. Decorso, M.; Anson, D.; Newby, R.; Wenglarz, R.; Wright, I.G. *Int. Gas Turbine and Aeroengine Congress*, Birmingham, UK, ASME Paper 96-GT-76, 1996.

26. Wright, I.G.; Leyens, C.; Pint, B.A. *in* Proc. ASME TURBOEXPO 2000, Munich, Germany, ASME Paper 2000-GT-0019, 2000.

27. Simms, N.J.; Encinas-Oropesa, A.; Kilgallon, P.J.; Oakey, J.E. Performance of Gas Turbine Materials in 'Dirty Fuel' Environments, to be published in *Materials for Power Engineering*, Liege, (Oct.2002).

28. Leyens, C.; Wright, I.G.; Pint, B.A. Hot Corrosion of Nickel-based Alloys by Alkali-containing Sulfate Deposits, *5th International Symposium on High Temperature Corrosion and Protection of Materials*, Les Embiez, France (May 2000).

29. Simms, N.J.; Smith, P.J.; Encinas-Oropesa, A.; Ryder, S.; Nicholls, J.R.; Oakey, J.E. in *Lifetime Modelling of High Temperature Corrosion Processes*, Eds M. Schütze *et al*, EFC No. 34 (Maney Publishing, London), p. 246260, 2001.

30. Simms, N.J.; Nicholls, J.R.; Oakey, J.E. in *Materials Science Forum*, v. 369-372, p. 833-840, 2001.

chapter 2

Technology Review of Modern Gas Turbine Inlet Filtration Systems

Melissa Wilcox[1] Rainer Kurz[2] and Klaus Brun[1]

[1]Machinery, Southwest Research Institute, San Antonio, Tx, USA
[2]Systems Analysis, Solar Turbines Incorporated, San Diego, CA, USA

ABSTRACT

An inlet air filtration system is essential for the successful operation of a gas turbine. The filtration system protects the gas turbine from harmful debris in the ambient air, which can lead to issues such as FOD, erosion, fouling, and corrosion. These issues if not addressed will result in a shorter operational life and reduced performance of the gas turbine. Modern day filtration systems are comprised of multiple filtration stages. Each stage is selected based on the local operating environment and the performance goals for the gas turbine. Selection of these systems can be a challenging task. This paper provides a review of the considerations for selecting an inlet filtration system by covering (1) the characteristics of filters and filter systems, (2) a review of the many types of filters, (3) a detailed look at the different environments where the gas turbine can operate, (4) a process for evaluating the site where the gas turbine will be or is installed, and (5) a method to compare various filter system options with life cycle cost analysis.

INTRODUCTION

Gas turbines ingest a large amount of ambient air during operation. Because of this, the quality of the air entering the turbine is a significant factor in the performance and life of the gas turbine. A filtration system is used to control the quality of the air by removing harmful contaminants that are present. The selection of the filtration system can be a daunting task, because there are many factors to consider. The system should be selected based on the operational philosophy and goals for the turbine, the contaminants present in the ambient air, and expected changes in the contaminants in the future due to temporary emission sources or seasonal changes. This paper outlines the primary considerations for selecting and installing a gas turbine inlet filtration system. First, the consequences that can occur due to improper inlet filtration are reviewed, then the different characteristics are discussed, after this the components of a filtration system and considerations for the operating environment are outlined, and lastly, a procedure for quantitatively comparing inlet filtration system options is provided.

CONSEQUENCES OF POOR INLET FILTRATION

When the quality of the air entering the gas turbine is not well controlled, there are several consequences which can occur. Some of the most common degradation mechanisms are reviewed below including erosion, fouling, and corrosion.

Erosion

Erosion occurs when solid or liquid particles approximately 10 μm and larger impact rotating or stationary surfaces in the gas turbine. The particles will impact the surface and remove tiny particles of metal which eventually lead to changes in the geometry of the surface. This change in geometry causes deviations in the air flow path, roughening of smooth surfaces, alteration of clearances, and reduction of cross-sectional areas of metal components possibly in high stressed regions. Erosion is a non-reversible process; therefore, the gas turbine components must be replaced in order to regain their original condition. However, particles 10 μm and larger are easily removed by commercial filters [1–3].

Fouling

Fouling of compressor blades is an important mechanism leading to performance deterioration in gas turbines over time. Fouling is caused by the adherence of particles to airfoils and annulus surfaces. Particles that cause fouling are typically smaller than 2 to 10 μm. Smoke, oil mists, carbon, and sea salts are common examples. Fouling can be controlled by an appropriate air filtration system and often reversed to some degree by detergent washing of components. The adherence is impacted by oil or water mists. The result is a build-up of material that causes increased surface roughness and to some degree changes the shape of the airfoil (if the material build up forms thicker layers of deposits). Fouling in turn causes a decrease in the performance of the gas turbine.

Commercial filters can remove the majority of particles that cause fouling. But there are several submicron particles that are difficult to remove from the flow stream. The build-up of particles not removed by the inlet filtration system is removed with the use of compressor washing. This process recovers a larger portion of the compressor performance but cannot bring the gas turbine back to its original condition [1–5].

Corrosion

When chemically reactive particles adhere to surfaces in the gas turbine, corrosion can occur. Corrosion that occurs in the compressor section is referred to as "cold corrosion" and is due to wet deposits of salts, acid, and aggressive gases such as chlorine and sulfides. Corrosion in the combustor and turbine sections is called "hot corrosion." It is also referred to as a high temperature corrosion. Hot corrosion requires the interaction of the metal surface with another chemical substance at elevated temperatures. Hot corrosion is a form of accelerated oxidation that is produced by the chemical reaction between a component and molten salts deposited on its surface. Hot corrosion comprises a complex series of chemical reactions, making corrosion rates very difficult to predict. It is the accelerated oxidation of alloys caused by the deposit of salts (e.g., Na_2SO_4). Type I or high temperature hot corrosion occurs at a temperature range of 1346 to 1742°F (730 to 950°C). Type II or low temperature hot corrosion occurs at a temperature range of 1022 to 1346°F (550 to 730°C). Some of the more common forms of hot corrosion are sulfidation, nitridation, chlorination, carburization, and vanadium, potassium, and lead hot corrosion. Sulfidation Hot Corrosion requires the interaction of the metal surface with sodium sulfate or potassium sulfate, salts that can form in gas turbines from the reaction of sulfur oxides, water, and sodium chloride (table salt) or potassium chloride, respectively. It is usually divided into Type I and Type II hot corrosion, and Type I hot corrosion takes place above the melting temperature of sodium sulfate (1623°F (884°C)), while Type II occurs below this temperature. Hot corrosion is caused by the diffusion of sulfur from the molten sodium sulfate into the metal substrate which prevents the formation of the protective oxidation film and results in rapid removal of surface metal. One should note that for hot corrosion to occur both sulfur and salt (e.g., sodium chloride or potassium chloride or chloride) have to be present in the very hot gas stream in and downstream of the combustor. Sulfur and salt can come from the inlet air, from the fuel, or water (if water is injected). The potassium hot corrosion mechanism is similar to sulfidation but is less frequently observed in gas turbines, unless the fuel contains significant quantities of potassium.

Corrosion is a nonreversible degradation mechanism. Therefore, corroded components must be replaced in order to regain the original gas turbine performance. Corrosion also initiates or advances other damage mechanisms in the gas turbine. For example, corrosion can intrude into cracks or other material defects and accelerates crack propagation [1–3].

FILTRATION CHARACTERISTICS

Filtration Mechanisms

Filters in the filtration system use many different mechanisms to remove contaminants from the air. The filter media, fiber size, packing density of the media, particle size, and electrostatic charge influence how the filter removes contaminants. Each filter typically has various different mechanisms working together to remove the contaminants. Four filtration mechanisms are shown in Figure 1.

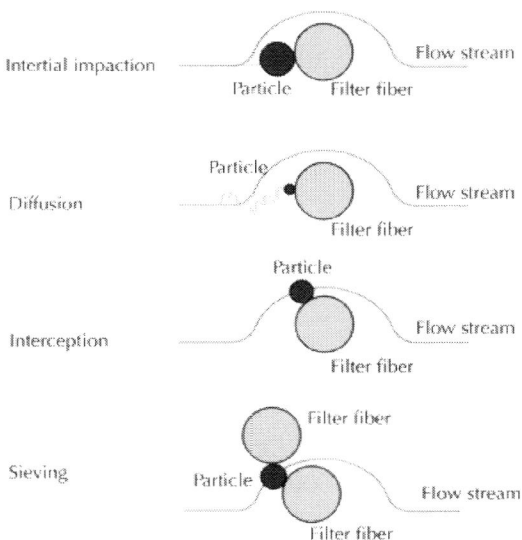

Figure 1: Common filtration mechanism [29].

The first filtration mechanism is inertial impaction. This type of filtration is applicable to particles larger than 1 µm in diameter. The inertia of the large heavy particles in the flow stream causes the particles to continue on a straight path as the flow stream moves around a filter fiber. The particulate then impacts and is attached to the filter media and held in place as shown in the top picture of Figure 1. This type of filtration mechanism is effective in high-velocity filtration systems.

The next filtration mechanism, diffusion, is effective for very small particles typically less than 0.5 µm in size. Effectiveness increases with lower flow velocities. Small particles interact with nearby particles and gas molecules. Especially in turbulent flow, the path of small particles fluctuates randomly about the main stream flow. As these particles diffuse in the flow stream, they collide with the fiber and are captured. The smaller a particle and the lower the flow rate through the filter media leads to a higher probability that the particle will be captured.

The next two filtration mechanisms are the most well-known: interception and sieving. Interception occurs with medium-sized particles that are not large enough to leave the flow path due to inertia or not small enough to diffuse. The particles will follow the flow stream where they will touch a fiber in the filter media and be trapped and held. Sieving is the situation where the space between the filter fibers is smaller than the particle itself, which causes the particle to be captured and contained.

Another mechanism not shown in Figure 1 is electrostatic charge. This type of filtration is effective for particles in the 0.01 to 1 µm size range (Figure 2). The filter works through the attraction of particles to a charged filter. In gas turbine applications, this charge is applied to the filter before installation as a result of the manufacturing process. Filters always lose their electrostatic charge over time because the particles captured on their surface occupy charged surface area, therefore neutralizing their electrostatic charge. As the charge is lost, the filter efficiency for small particles will decrease. On the other hand, as the filter is loaded, the filtration efficiency increases, thus counteracting the effect of lost charge to some extent. This will offset some of the loss of filtration efficiency due to the lost charge. Figure 2 shows a comparison of a filter's total efficiency based on the various filtration mechanisms that are applied. The figure shows the difference between the filter's efficiency curve before and after the charge is lost. The performance of the filter should be based on the discharged condition [6, 7].

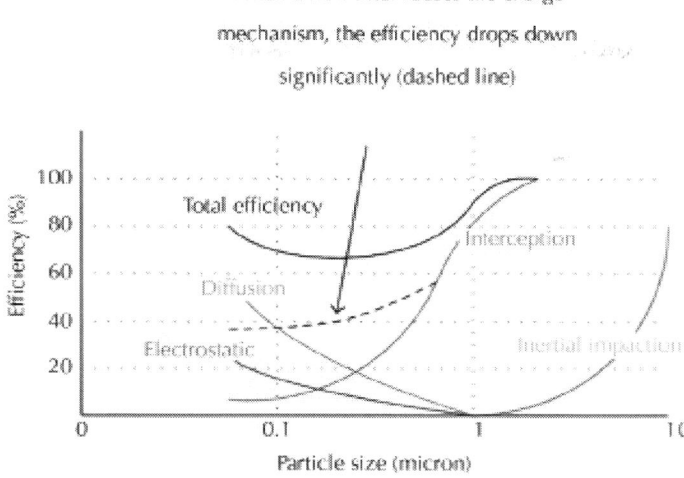

When a new filter looses the charge
mechanism, the efficiency drops down
significantly (dashed line)

Figure 2: Combination of filtration mechanisms to obtain filter efficiency at various particle sizes [29].

Filter Efficiency and Classification

Filter efficiency is a broad term. In general, the filter efficiency is the ratio of the weight, volume, area, or number of particles captured in the filter to the weight, volume, area, or number of the particles entering the filter, respectively. A general efficiency calculation is shown in (1), where W is the variable for which efficiency is being calculated. The efficiency can be expressed in several ways: maximum, minimum, or average lifetime value. Many filters have poor performance against small particles at the beginning of their lives, but as the filter media becomes loaded with particles, it is able to catch smaller particles. In this case, the average efficiency would actually be higher than the initial efficiency. Some of the filters will never reach the quoted maximum efficiency before they are replaced

$$\eta = W_{entering} - W_{leaving} / W_{entering} * 100\% \tag{1}$$

Filter efficiency is a trade-off against the pressure loss across the filter. Normally, the filtration system pressure loss will increase with an increase in filtration efficiency. As filters become more efficient, less dust penetrates through them. Also, the air flow path is more constricted with higher efficiency filters. This leads to higher pressure loss. Filter engineers must determine the acceptable pressure loss and efficiency for their application. Studies have shown that a higher pressure loss due to using a high efficiency filter has a lower effect on gas turbine power degradation than poor inlet air quality.

The efficiency of a filter cannot be stated as a general characteristic. The filter efficiencies vary with particle size, typically being lower for small particles and higher for large particles. They also vary with operational velocity. Filters designed for medium and low velocities will have a poor performance at higher velocities and vice versa. Therefore, a particle size range and flow velocity must be associated with the stated efficiency. For example, a filter may have 95 percent filtration efficiency for particles greater than 5 µm at a volumetric flowrate of 3000 cfm (5097 m³/h), but the efficiency could be reduced to less than 70 percent for particles less than 5 µm or at a volumetric flowrate of 4000 cfm (6796 m³/h).

Filters are rated for performance based on standards established in the United States of America and Europe. These filter ratings are based on the results of standard performance tests. In the United States, ASHRAE standard 52.2:2007 outlines the requirements for performance tests and the methodology to calculate the efficiencies. In this standard, the efficiencies are determined for various ranges of particles sizes. The filter is given a Minimum Efficiency Reporting Value (MERV) rating based on its performance on the particle size ranges (particle count efficiency) and the weight arrestance (weight efficiency). The weight arrestance is a comparison of the weight of the dust penetrating the filter to the dust feed into the flow stream. In this standard, a filter with a MERV of 10 will have 50 to 65 percent minimum efficiency for particles 1 to 3 µm in size and greater than 85 percent for particles 3 to 10 µm in size [8, 9].

The European standards used to determine performance are EN 779:2002 and EN 1822:2009. EN 779:2002 is used to rate coarse and fine efficiency filters. EN 1822:2009 presents a methodology for determining the performance of high efficiencies filters: Efficient

Particulate Air filters (EPA), High Efficiency Particulate Air filter (HEPA), and Ultra Low Particulate Air filter (ULPA). In EN 779:2002, the performance is found with average separation efficiency, which is an average of the removal efficiency of 0.3 μm particles at four test flowrates (particle count efficiency) for fine filters and with an average arrestance (weight efficiency) for coarse particle filters. This standard rates the filters with a letter and number designation: G1 to G4 (coarse filters) and F5 to F9 (fine filters). Filter performance is determined by the Most Penetrating Particle Size efficiency (MPPS) in EN 1822:2009. The MPPS is defined as the particle size, which has the minimum filtration efficiency or maximum penetration during the filter testing. The particle sizes tested range from 0.15 to 0.3 μm. The filter efficiency is calculated based on particle count. These filters are given a rating of E10 to E12 for EPA-type filters, H13 to H14 for HEPA-type filters, and U15 to U17 for ULPA filters. Table 1 gives a general overview of the efficiencies for each filter rating and a comparison of the filter ratings between American and European standards [10–16].

Table 1: Summary of filter classification for ASHRAE 52.2:2007 [10], EN 779:2002 [11], and EN 1822:2009 [12–16]

	ASHRAE 52.2:2007 [10]				EN 779:2002 [11]		EN 1822:2009 [12–16]	
ASHRAE filter class	Average particles size efficiencies in micron (%)			EN filter class	Average separation efficiency (A_m)	Average separation efficiency (E_m)	Total filtration separation efficiency (%)	Local filtration separation efficiency (%)
	E_1	E_2	E_3					
MERV	0.3–1.0	1.0–3.0	3.0–10.0					
1			<20	G1	$50 \leq A_m < 65$			
2			<20					
3			<20	G2	$65 \leq A_m < 80$			
4			<20					
5			20–35	G3	$80 \leq A_m < 90$			
6			35–50					
7			50–70	G4	$90 \leq A_m$			
8			>70					
9		<50	>85	F5		$40 \leq E_m < 60$		
10		50–65	>85					

11		65–80	>85	F6		$60 \leq E_m < 80$		
12		>80	>90					
13	<75	>90	>90	F7		$80 \leq E_m < 90$		
14	75–85	>90	>90	F8		$90 \leq E_m < 95$		
15	85–95	>90	>90	F9		$95 \leq E_m$		
				E10			85	
16	>95	>95	>95	E11			95	
				E12			99.5	
17				H13			99.95	99.75
18								
19				H14			99.995	99.975
20								
				U15			99.9995	99.9975
				U16			99.99995	99.99975
				U17			99.999995	99.9999

Note: Correlations between ASHRAE and EN standard are approximate.

If an engine ingests 220 lb/year (100 kg/year) of contaminants if there were no filtration system in a typical offshore application, an F5 filter would reduce this to about 46 lb/year (21 kg/year), an F6 filter to 13 lb/year (6 kg/year), an F7/E10 filter system to 0.44 lb/year (0.20 kg/year), and an F7/F9/E10 system to as little as 0.11 lb/year (0.05 kg/year). This indicates two conclusions: while large particles have an impact on fouling degradation, a significant amount is due to the finer particles. The overall contaminant ingestion can be influenced by several orders of magnitude by using an appropriate air filtration system. Also, with filtration systems of this type, there are virtually no particles larger than a few microns entering the engine [17].

Filter Pressure Loss

As mentioned above, a higher pressure loss occurs with a more efficient filter due to air flow restrictions. Pressure loss has a direct impact on the gas turbine performance, as it causes a reduction in compressor inlet pressure. For the compressor to overcome the inlet system losses, the gas turbine will consume more fuel, and it also has a reduced power output. As the pressure loss increases the power decreases and the heat rate increases linearly. A 0.2 in H2O (50 P_a) reduction of pressure loss can result in a 0.1 percent improvement in power output. Typical pressure losses on inlet filtration systems can range from 2 to 6 in H2O (500 to 1500 P_a) [18].

The filter's performance needs to be assessed for the full pressure loss range over its life, not just when it is new. The pressure loss will increase over the lifetime of the filter. Therefore, one can expect a lower gas turbine performance over the life of the filter, or filters have to be changed or cleaned periodically in order to maintain a low pressure loss. The change of pressure loss over time is highly dependent upon the filter selection and the type and amount of contaminants experienced.

Filter Loading (Surface or Depth)

During operation as the filter collects particles, it is slowly loaded until it reaches a "full" state. This state is usually defined as the filter reaching a specified pressure loss, or when the end of maintenance interval. Filters are loaded in two different ways: surface and depth loading.

Depth loading is the type of filtration where the particles are captured inside of the filter material. To regain the original pressure loss or condition, the filter must be replaced.

The other type of filter is a surface-loaded filter. With this type of loading, the particles collect on the surface of the filter. Few of the particles may penetrate the fiber material, but not enough to call for a replacement of the filter. Surface loaded filters are most commonly used in, but not restricted to, self-cleaning systems, because the dust can easily be removed with pulses of air once the filter differential pressure reaches a certain level. Once the filter is cleaned, the pressure loss across the filter will be close to its original condition. The surface loaded filter's efficiency actually increases as the surface is loaded with dust, because a dust cake develops on the surface of the media, creating an additional filtration layer, and also decreases the amount of available flow area in the filter media [19, 20].

Face Velocity

Filtration systems are distinctively classified as high, medium, or low velocity systems. The velocity of the filtration system is defined as the actual volumetric air flow divided by the total filter face area. Low velocity systems have air flow at less than 500 ft/min (2.54 m/s) at the filter face. Medium velocities are in the range of 610 to 680 ft/min (3.1 to 3.45 m/s). High velocity systems have air flows at the filter face in excess of 780 ft/min (4 m/s).

High Velocity Systems

Historically, high velocity systems are used on marine vessels and offshore platforms where space and weight are premiums. However, today, low, medium, and high velocity systems are found on marine and offshore applications. High velocity systems have the advantages of reduced size (cross-sectional area), weight, and initial cost. Filter efficiencies for small particles are significantly lower than those of lower velocity systems, and dust holding capacities are lower.

High velocity systems typically use vane separators upstream, and often also downstream, of the filter media to remove water from the air stream. For the vanes to work effectively, higher flow velocities are necessary. Ultimately, this type of system requires more filter replacements when compared to the lower velocity system of similar performance [21, 22].

Low Velocity Systems

Low velocity systems are the standard on land-based applications; however, high velocity systems are also used in some coastal applications. The low velocity systems are characterized by large inlet surface areas, large filter housings, and usually multiple stages of filters. The two or three stage filters provide an advantage over high velocity systems, because they have a high efficiency filter stage as the final stage to remove many small particles (especially salt) below 1 µm. Recently developed filter media can also keep water from penetrating the media, and thus entering the gas turbine. The lower velocity also provides a lower pressure loss or higher filtration efficiency. Using prefilters to remove the majority of the particles, the life of the high efficiency filters is extended. Overall, low velocity systems can be more effective at reducing the mass of contaminants that enter a system, thus extending the water wash intervals for the engine (Figure 3).

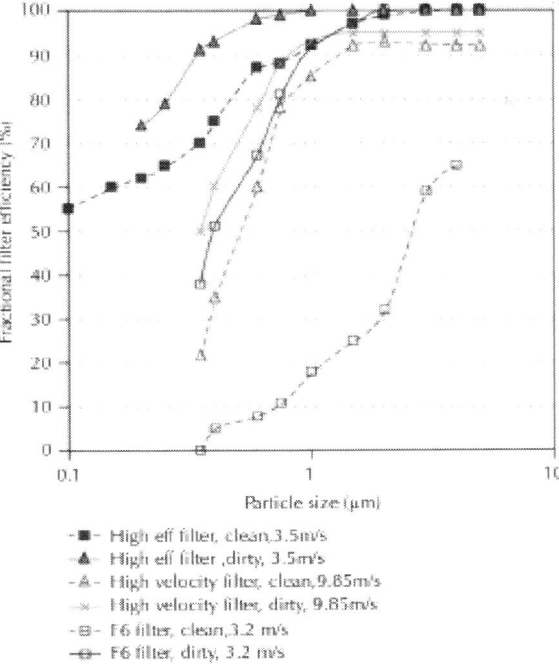

Particle size (μm)

- ■ - High eff filter, clean, 3.5m/s
- ▲ - High eff filter, dirty, 3.5m/s
- △ - High velocity filter, clean, 9.85m/s
- × - High velocity filter, dirty, 9.85m/s
- □ - F6 filter, clean, 3.2 m/s
- ⊖ - F6 filter, dirty, 3.2 m/s

Figure 3: Comparison of fractional efficiency for filter elements from different suppliers and different face velocities in new and dirty conditions [25].

Water and Salt Effects

Many environments where gas turbines operate will have wet ambient conditions. This could be in a tropical environment where it rains a significant amount of time or coastal location with ocean or lake mist. Table 2 is a list of the different types of moisture that can be experienced together with their particle size. The difference between filter operation in wet and dry conditions can be significant. In some cases, the pressure loss across a filter can increase significantly even with a little moisture. This is true for cellulose fiber filters which swell when they are wet. These filters will also retain the moisture that can lead to long periods of time when the pressure loss across the filter is elevated.

Table 2: Different types of moisture experience in inlet filtration systems [2, 23]

Description	Liquid Size (μm)
Humidity	vapor form
Smog (more smoke than humidity)	0.01 to 2
Cooling tower aerosols	1 to 50
Water mist	1 to 50
Clouds and fog	2 to 150
Water spray (ship wake, ocean spray)	10 to 500
Drizzle	50 to 400
Rain	400 to 1000

Salt can have a direct effect on the life of a gas turbine if not removed properly. It is often carried into the engine dissolved in water spray. Salt can lead to fouling and corrosion. Gas turbine manufacturers usually recommend stringent criteria on the amount of salt which can be allowed to enter the gas turbine (less than 0.01 ppm). In coastal environments, the airborne salt can easily range from 0.05 to 0.5 ppm on a typical day. If the filtration system is not equipped to handle the salt, it can enter the compressor and the hot section of the gas turbine. Salt is present in the air, either as salt dust or dissolved in seawater, and contains sodium chloride, magnesium chloride, and calcium sulfate. Salt may also come from localized sources such as a dry salt bed [2, 23]. The salt on compressor blades must be removed through water washing methods or direct scrubbing of the blades.

COMPONENTS OF A FILTRATION SYSTEM

In order to protect the gas turbine from the variety of contaminants present in the ambient air, several filtration devices are used. Each of the devices used in modern filtration systems are discussed below.

Weather Protection and Trash Screens

Weather louvers or hoods and trash screens are the most simplistic of the filtration mechanisms, but they are important in order to reduce the amount of moisture and solid contaminants, which enter the main filtration system. These are not classified as filters, but they are part of the filtration system and provide assistance in removal of large objects or contaminants carried in the flow stream.

Weather hoods are sheet metal coverings on the entrance of the filtration system (see Figure 4). The opening of the hood is pointed downward so the ambient air must turn upwards to flow into the inlet filtration system. The turning of the air is effective at minimizing rain and snow penetration. Weather hoods and louvers are used on the majority of inlet filtration systems, and they are essential for systems in areas with large amounts of rainfall or snow. Weather hoods or another comparable weather protection system are strongly recommended for all systems with high efficiency filter.

Figure 4: Weather Hood on inlet filtration system [34].

After the weather hood is a series of turning vanes called weather louvers, which redirect the air so that it must turn. The weather louvers are also effective at minimizing water and snow penetration. After the weather hood or louver is a trash or insect screen. Trash screens capture large pieces of paper, cardboard, bags, and other objects. The screens also deflect birds, leaves, and insects. Screens that are installed specifically for preventing insects entering the filtration system are referred to as insect screens. These screens will have a finer grid than trash screens. Weather hoods, louvers, trash screens, and insect screens are used on the majority of filtration systems due to their inexpensive cost and construction, and negligible pressure loss [9].

Anti-Icing Protection

Anti-icing protection is used in climates with freezing weather. Freezing climates with rain or snow can cause icing of inlet components, which can result in physical damage to inlet ducts or to the gas turbine compressor. This ice can also affect the performance of the gas turbine. If ice forms on filter elements, then ice on those filters will be blocking the flow path, which will cause the velocity at the other filters to increase. This causes a decrease in filtration efficiency. Also, the filter elements with ice can be damaged. Figure 5 shows an example of ice formation on filters due to cooling tower drift.

Figure 5: Pulse cartridge filters with frost build up due to cooling tower drift [29].

Heaters, compressor bleed air, or self-cleaning filters are often used in the inlet system in frigid environments to prevent the build-up of ice on the inlet bell mouth or filter elements. It should be noted that any location in the inlet system (even past the filter system) that creates a pressure drop can potentially have ice formation. In some plant operations, the Inlet Guide Vanes (IGVs) are used for flow control at part load operations. If the IGVs are partially closed, then under the right weather conditions, ice build can occur at this location. In order to prevent ice build-up in this situation, it may be necessary to limit the closer of the IGVs or have provisions to heat the inlet air to avoid ice formation.

Inertial Separators

Inertial separation takes advantage of the physical principles of momentum, gravity, centrifugal forces, and impingement, and the physical difference between phases to cause particles to be moved out of the gas stream in such a way that they can be carried off or drained. The higher momentum of the dust or water particles contained in the air stream causes them to travel forward, while the air can be diverted to side ports and exit by a different path than the dust. There are many types of inertial separators, but the ones commonly used with gas turbine inlet filtration are vane and cyclone separators [7, 9].

Moisture Coalescers

In environments with high concentration of liquid moisture in the air, coalescers are required in order to remove the liquid moisture. The coalescer works by catching the small water droplets in its fibers. As the particles are captured, they combine with other particles to make larger water droplets. Coalescers are designed to allow the droplets to either drain down the filter or be released back into the flow stream. If the larger drops are released, then they are captured downstream by a separator. Figure 6 shows an example of how the droplet size distribution changes across the coalescer which releases the droplets [7, 9].

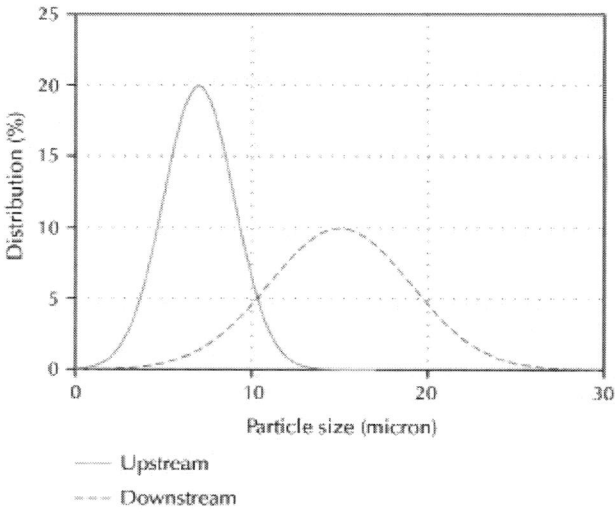

Figure 6: Coalescer droplet formation distribution [29].

Prefilters

The air has a mixture of large and small particles. If a one-stage high efficiency filter is used, the build-up of large and small solid particles can quickly lead to increased pressure loss and filter loading. Prefilters are used to increase the life of the downstream high efficiency filter by capturing the larger solid particles. Therefore, the high efficiency filter only has to remove the smaller particles from the air stream which increases the filter life. Prefilters normally capture solid particles greater than 10 µm, but some prefilters will also capture the solid particles in the 2 to 5 µm size range. These filters usually consist of large diameter synthetic fiber in a disposable frame structure. Bag filters are also commonly used for prefilters. These offer higher surface area that reduces the pressure loss across the filter [7, 9]. In many installations, the prefilters can be exchanged without having to shut the engine down.

High-Efficiency Filters

As discussed above, there are filters for removing larger solid particles, which prevent erosion and FOD. Smaller particles which lead to corrosion, fouling, and cooling passage plugging, are removed with high efficiency filters. These types of filters have average separations greater than 80 percent. Three common types of high efficiency filters are EPA, HEPA, and ULPA. EPA and HEPA filters are defined as having a minimum efficiency of 85 percent and 99.95 percent, respectively, for all particles greater than or equal to 0.3 µm. ULPA filters have a minimum efficiency of 99.9995 percent for particles the same size or larger than 0.12 µm [11–16]. Often, these names are used loosely with discussion of high efficiency filtration. However, the majority of the high efficiency filters used in gas turbine inlet filtration do not meet these requirements.

The high-efficiency filters used with gas turbines have pleated media that increase the surface area. In order to achieve the high filtration efficiency, the flow through the filter fiber is highly restricted which creates a high pressure loss, unless the face velocity is kept low. The pleats help reduce this pressure loss. Initial pressure loss on high efficiency filters can be up to 1 inH2O (250 P_a) with a final pressure loss in the range of 2.5 in H2O (625 P_a) for rectangular filters and 4 inH2O (2000 P_a) for cartridge filters. The life of the filters is highly influenced by other forms of filtration upstream. If there are stages of filtration to remove larger solid articles and liquid moisture, then these filters will have a longer life. Minimal filtration before high efficiency filters will lead to more frequent replacement or cleaning. High efficiency filters are rated under various standards. The majority of filters used in gas turbines are not classified as EPA, HEPA, or ULPA. The filters used in gas turbines are rated with ASHRAE 52.2:2007 and EN 779:2002.

There are many different constructions of high efficiency type filters: rectangular, cylindrical/cartridge, and bag filters. The rectangular high efficiency filters are constructed by folding a continuous sheet of media into closely spaced pleats in a rectangular rigid frame. Rectangular filters are depth loaded; therefore, once they reach the maximum allowable pressure loss, they should be replaced. Two examples of rectangular high efficiency filters are shown in Figure 7. High efficiency filters can also be made from media that do not allow water to seep through the filter media.

Figure 7: Rectangular high-efficiency filters [35, 36].

Cartridge filters are also made up of closely spaced pleats, but they are in a circular fashion (Figure 8). Air flows radially into the cartridge. They are installed in a horizontal or vertical fashion (hanging downward). These types of filters can be depth or surface loaded. The surface loaded filters are commonly used with a self-cleaning system, but not all of them are designed for self-cleaning. Cartridge filters used in self-cleaning systems require a more robust structural design in order to protect the filter fiber media during the reverse air pulses. The more common structural support is a wire cage around the pleated media on the inside and outside of the filter. The filters shown in Figure 8 are not designed for a self-cleaning system since there are no structural supports on the outside of the filter. Self-cleaning filtration systems are discussed in the next section [7, 9].

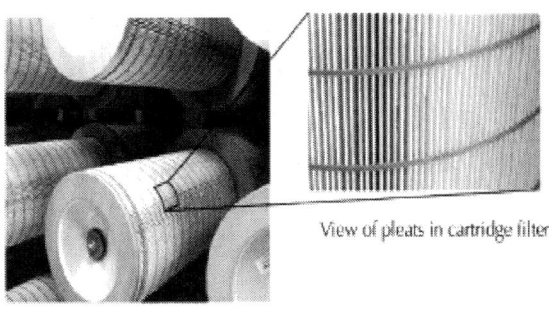

View of pleats in cartridge filter

High-efficiency cartridge filters

Figure 8: High-efficiency cartridge filters [29].

Self-Cleaning Filters

All of the filters with fiber-type media previously discussed are required to be replaced once they reach the end of their usable life. In some environments, the amount of contaminants can be excessive to the point where the filters previously discussed would have to be replaced frequently to meet the filtration demand. A prime example of one of these environments is a desert with sand storms. In the 1970s, the self-cleaning filtration system was developed for the Middle East where gas turbines are subject to frequent sand storms. Since then, this system has been continually developed and utilized for gas turbine inlet air filtration.

The self-cleaning system operates primarily with surface-loaded high-efficiency cartridge filters. The surface loading allows for easy removal of the dust, which has accumulated with reverse pulses of air (Figure 9). The pressure loss across each filter is continuously monitored. Once the pressure loss reaches a certain level, the filter is cleaned with air pulses. The pressure of the air pulses ranges from 80 to 100 psig (5.5 to 6.9 barg). The reverse jet of compressed air (or pulse) occurs for a length of time between 100 and 200 ms. to avoid disturbing the flow and to limit the need for compressed air, the system typically only pulses 10 percent of the elements at a given time. With this type of cleaning, the filter can be brought back to near the original condition [19, 24].

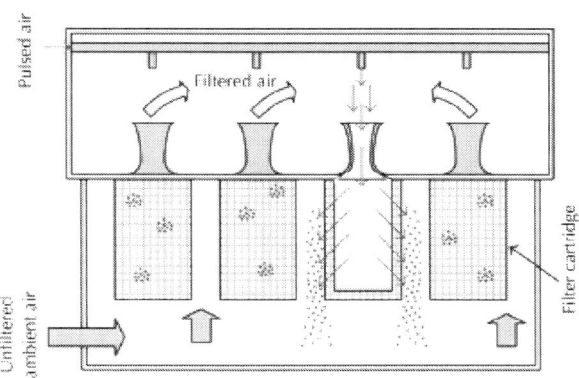

Figure 9: Example of operation of an updraft self-cleaning filters [29].

Staged Filtration

Any gas turbine application typically needs more than one type of filter, and there are no "universal filters" that will serve all needs. Therefore, two-stage or three-stage filtration systems are used. In these designs, a prefilter or weather louver can be used first to remove erosive contaminants, rain, and snow. The second may be a low-to-medium-performance filter selected for the type of finer-sized particles present or a coalescer to remove liquids. The third filter is usually a high-performance filter to remove smaller particles less than 2 μm in size from the air. Figure 10 shows a generalized view of a filtration arrangement. This arrangement is not correct for all cases due to the fact that the filter stages are highly influenced by the environment they are operating in.

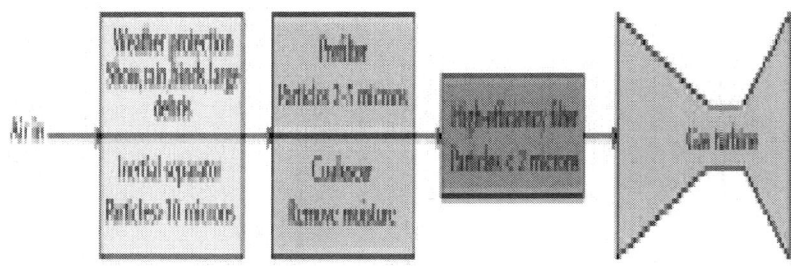

Figure 10: Multistage filtration system.

OPERATING ENVIRONMENT

The selection of the inlet filtration system should be primarily dependent on the environment where it operates. This includes the contaminants in the ambient air from surrounding vegetation, weather events, local emissions, temporary emissions, and seasonal changes. Several different common environments with their typical contaminants are reviewed below.

Coastal, Marine, or Offshore

Gas turbines operating near or on the ocean are classified as being in a coastal, marine, or offshore environment. The gas turbine is in a coastal environment when the gas turbine is installed on land and within 10 miles (16 km) of the ocean. At distances from approximately 8 to 12 miles (13 to 19 km) from the shoreline, the salt concentration in the air drops to natural background levels for an environment far away from the ocean [23]. The offshore and marine environments are defined as being in the middle of the ocean. The gas turbine is considered in an offshore environment when it is located at least 100 ft (30.5 m) off of the ocean surface. Gas turbines located below 100 ft (30.5 m) are considered to be in a marine environment.

The primary contaminant that is a concern in the coastal, marine, and offshore environments is salt. Salt as previously discussed can lead to fouling and corrosion. Salt is prevalent in these environments due to the sea water. In coastal environments, it is present as dry contaminants (areas with lower humidity), sticky contaminants (humidity between 40 and 70 percent), or as liquid aerosols (humidity greater than 70 percent) [2]. In the offshore environment, salt is usually present in the sticky particle or liquid state. The marine environment (closest to the ocean surface) has salt present in the liquid aerosol state. In all these environments, the amount of salt depends on the wind speed and direction and the elevation of the gas turbine.

Coastal environments also have land-based contaminants that must be removed from the air. These will be discussed in more detail below. Offshore environments have industrial contaminants such as exhaust fumes, by-products of maintenance (such as dust from grit blasting), and unburned hydrocarbons from flares. Many of these particles are on the submicron size; therefore, high efficiency filtration is often employed. The marine environment most often does not have as many additional contaminants to remove from the air. However, when a ship is near a coast, land-based contaminants may be present. In addition, icing in colder environments is often a concern. Icing can also be an issue in offshore and coastal environments.

The filtration system for a coastal environment is similar to that of a land-based environment, which will be discussed later. However, in coastal environment it is important to have mist eliminators for water and salt water removal and high efficiency filtration for salt removal. The filtration system in an offshore environment is similar to a coastal filtration system but may have increased air velocity due to size and weight limitations. The filtration system on a marine vessel is most commonly comprised of a vane-coalescer-vane system. This system has two vane axial separators with a coalescer in between them. This system is a high velocity system that is designed for removing salt water. It has limited solid particle removal capability [2, 23, 25, and 26].

Land-Based Environment

The land-based environment is very diverse. It can be classified in many different ways depending on weather patterns, vegetation, and local emission sources. Several land-based environments are described below.

Desert

The desert is classified as an area with a dry and hot climate. Large amount of dust is present and there is little vegetation. Sand storms are common and can quickly load filters to their maximum dust holding capacity. The main regions of the world which can be characterized by desert like environments are across the Sahara desert in Africa, the Middle East, and parts of Asia. However, small localized areas with high dust concentrations do exist. These can include gas turbines installed near quarries, dried lakebeds, loess, industrial areas, dirt tracks, dry agricultural land, and construction sites. There are three typical conditions that exist in the desert: clean air, dust haze, and sand storms. Dust is the main contaminant in the desert for these conditions. This can be sand or other fined grained material such as desert pavement. Desert pavement is the layer of large stones left on the floor of the desert. While these stones are not harmful in their solid state, they can easily be broken by human or animal traffic and crumbled into fine particles. These particles can range from large (500 µm) to very fine (submicron size). Due to the lack of vegetation and protection of the ground dust from the wind, more dust can be lofted into the air than in other environments. This leads to a high concentration of dust.

The filtration systems in deserts are usually solely designed for dust removal. However, some desert locations experience periods of dense fog and high humidity. This is especially true for deserts near a coastal region. The moisture can collect on the surface of cartridge filters on self-cleaning systems and cause the dirt to form a cake on the filter. This cake of dust can significantly reduce the effectiveness of filtration and pulse cleaning. If fog and high humidity are present at the desert-type site, then this should be considered for the filtration system.

Dust loads in the desert can range from mild (low wind) to fairly high (dust storms). Conventional non-self-cleaning filtration systems can quickly become loaded and require frequent filter change outs. Also, high pressure losses can trigger a shutdown if they become excessive. In order to avoid the constant maintenance and labor required for changing filters out, a self-cleaning system is needed. Filtration systems without self-cleaning filters have proven to be more expensive due to the labor cost and maintenance required with filter replacements [26, 27].

Arctic

The arctic environment is characterized by freezing weather (below 32°F (0°C)) for an extended period of time. The location will not necessarily be classified as arctic for the entire year. It will have other land-based contaminants, which must be considered. However, the arctic seasons of the year will influence the design of the inlet filtration system.

Ice build-up is the primary concern in this environment during the cold months. Ice can form from the ingestion of snow or freezing rain and also due to the depression or cool humid air in the inlet system. Placement of the inlet of the filtration system, weather hoods with large openings (referred to as snow hoods), and self-cleaning filters is often adequate to protect against the ingestion of snow and freezing rain. To prevent the formation of ice from the depression of cool humid air requires an anti-icing system such as recirculated exhaust air or a compressor bleed.

In addition to ice, warm season contaminants must be considered for the design of the inlet filtration system. These contaminants can be similar to any of the other land-based environments discussed in this section [26].

Tropical

Tropical areas are characterized by hot climate, high humidity, monsoons, high winds, and insect swarms. Due to the extensive vegetation, there is not much erosion concern. It is considered a low-dust environment. The area has little seasonal variation with the exceptions of periods of intense rainfall. Typhoons, dust, insects, and the remoteness of systems in the tropics should be considered when choosing the correct system.

The main contaminants in this area are water (from rain), insects, and salt (if the location is near a shoreline). Dust is minimal, since the overgrown vegetation protects the ground dust from winds. Of course, there are always exceptions to this. If the gas turbine is installed in a construction site, then the dust levels will be higher than normal. Also, unpaved roads can contribute to the dust in the environment. Pollen can be an issue. Salt will be present in aerosol form due to the high humidity and moisture present.

The filtration systems for tropical environments are specifically built to handle large amounts of rain. Weather hoods are used as a primary defense. Extended area insect screens are used for blocking insects. These screens have a lower air velocity (in the range of 260 ft/min (1.3 m/s)), which allows the insects to move away from the screens. This prevents obstruction of the inlet air flow. This is followed by a mix of prefilters, coalescers, and vane separators. The water removal system must be designed in order to handle the highest expected water ingestion and prevent corrosion. If this is not done, then water will be able to travel farther downstream in the inlet filtration system. Any prefilters or high efficiency filters used should be selected to prevent water travel through the filter. If water is allowed to penetrate the filter, then it can absorb the capture soluble contaminants and transport them through the filter into the gas turbine. This can have detrimental effects if salt is being removed from the air stream. These filters should also be selected for the expected contaminants such as pollen and road dust [26, 28].

Rural

The rural countryside is a diverse environment. Depending upon where the gas turbine is located in this environment, it can be subjected to hot, dry climate, rain, snow, and fog throughout the year. The majority of the year there is a nonerosive environment with low dust concentrations in the range of 0.02 to 0.1 ppm (0.01 to 0.05 grains per 1000 ft^3 (28.3 m^3)). The area can be near a local forest or be near agricultural activities.

The contaminants in this environment vary depending on the season. Throughout the year, insects and airborne particulate will need to be filtered. If the gas turbine is installed near an agricultural area, then during plowing and harvesting season, the concentration of dust will increase. During plowing, insecticides and fertilizers will be airborne. At harvest, the particles or grains from cutting plants down will be lofted into the air. The particles that travel to the gas turbine are relatively small (less than 10 μm), unless strong winds are present to carry large particles. Gas turbines near forests may not have as high dust concentration. The foliage of the forest will protect the ground dust from being lifted by the wind. With the change in season, snow, rain, fog, pollen, airborne seeds, and insects will be present. This climate has one of the most diverse filtration requirements as compared to other environments.

These systems are typically comprised of three stages: weather hood, prefilter, and high-efficiency filter. The weather hood protects the filters farther downstream from rain and snow. They also minimize the amount of dust entering the filtration system. Insect screens are used, especially if insects are present in swarms during parts of the year. The prefilter is used to remove any erosive dust present in the air. The prefilter also protects the high efficiency filter from being overloaded too quickly. The high efficiency filter removes the smaller particles. If the gas turbine is installed near an agricultural area, the filter engineer may consider a self-cleaning system. This type of system would be beneficial during plowing or harvest season when the air has a high erosive dust concentration. A self-cleaning system can also be beneficial in an area with a dry, cold climate during the winter season. It can effectively prevent ice from forming on the filter elements and influencing the gas turbine operation [26].

Large City

Large cities can experience all the types of gas turbine degradation: corrosion, erosion, and fouling. Contaminants from many different sources ensure the requirement of a multistaged filtration system.

All different types of weather can occur throughout the year in a large city. The amount of contaminants varies throughout the season as discussed above for the rural countryside. One example is salt or grit that is laid down on icy roads during the winter. The city also has smog and pollution. These can also be seen in the countryside due to high winds, but are much more concentrated in the city. Some other considerations for large cities are noise issues and vandals.

The system has a multistage approach with specific filters installed for the local contaminants. Weather hoods are used the majority of the time due to the changing weather conditions with seasons. This protects the system from rain, snow, and windy conditions. The filtration system is composed of a prefilter and a high-efficiency filter. The prefilter removes the larger erosive particles. The high-efficiency filter is typically of the non-self-cleaning type with rectangular filters or cartridges filters. The self-cleaning systems are not used due to the sticky aerosols present in the air. If freezing conditions are expected, then an anti-icing system is included. Urban/industrial areas typically do not have airborne particulate concentrations that warrant the use of self-cleaning filtration systems, but self-cleaning systems are used successfully in these areas, when these are in regions of heavy snow and minimal sticky contaminants [26].

Industrial Area

Many gas turbines are installed in heavy industrial areas. These locations can be in any of the environments discussed above, but they have additional concerns. There are several emission sources in an industrial location, which contribute to the contaminants that must filtered out.

The most prevalent contaminant in industrial areas is contaminants from exhaust stacks. These can be in the form of particles, gases, and aerosols. Many of the particles emitted by the exhaust stack are in the submicron size range. These size particles are difficult to filter and can collect on compressor blades and cause fouling. The gases emitted in the exhaust can contain corrosive chemicals. For example, exhaust gases from fossil fuel plants has SO_x, which contains sulfur. Sulfur is one of the corrosive components that can lead to hot corrosion in the turbine section. Gas cannot be removed by mechanical filtration. Aerosols also present a challenge. These are typically on the submicron size and difficult to filter. Many of these aerosols are sticky, and when they are not removed by the filters, they stick to compressor blades, nozzles, and other surfaces. One example of this already mentioned in this guideline is the compressor blade fouling due to oil vapors.

Industrial locations can also experience contaminants that are not typically seen, unless near a localized source. Some examples of these are dust from mining operations, sawmills, foundries, and other industrial facilities. Also, if the gas turbine is near a petrochemical plant, the air may be contaminated with specific chemicals. These chemicals could be harmless, but they also could have corrosive properties.

One commonality between all industrial locations is that the inlet of the filtration system is subjected to the local plant emissions. This condition typically requires a more robust high-efficiency filtration system to remove fine particles that are entrained in the air. One way to reduce the amount of emissions that are ingested into the inlet is to direct the inlet air flow away from these emission sources. Several recommendations in regards to the inlet placement and site layout are discussed in a later section. Even so, there are still some emissions that are ingested by the turbine. Additional filter elements should be included in the filtration system to address these emission particles. For example, if the industrial location is near an open coal storage site, then the gas turbine should have prefilters and high-efficiency filters to remove the coal dust that is in the air.

One contaminant that is often in the air at industrial locations is sticky aerosols. These aerosols can be from oil vapors from lubrication systems or unburned hydrocarbons emitted from exhaust stacks. These aerosols are very difficult to remove from the air and often lead to blade fouling. High-efficiency filters should be used to minimize the aerosol's effect on the gas turbine, but a compressor washing scheme is needed to keep the compressor blades clean and to minimize the effects of fouling on gas turbine performance [26].

Temporary and Seasonal Contaminant Sources

In many of the applications discussed above, temporary or seasonal conditions are mentioned. As gas turbines become more advanced and more sensitive to the inlet air quality, it becomes more important to address these conditions.

In order to address seasonal changes, the expected conditions must first be defined. During the design phase, the air quality at the site where the gas turbine is going to be installed should be monitored for at least 1 year. This will provide the filter engineer with information about which contaminants they can expect in each season. Also, the filter engineer should map out any potential construction, agricultural, or dust-generating projects that will occur in the first 5 to 10 years of the life of the gas turbine. Combining the expected contaminants will allow the filter engineer to develop a more holistic approach to their inlet filtration.

Currently, the majority of the filtration systems installed have a fixed filtration system. The number of stages, types of filters, and level of filtration remain constant throughout the operation. If the future site for the gas turbine is expected to have high variability in the type of contaminants experienced (temporary or seasonal), the filter engineer may consider a filtration system which allows the use of many different filters. This would then allow the filtration system to be adapted to the current conditions.

Site Layout

The layout of the site where the gas turbine is installed can have a significant effect on the type and amount of contaminants that need to be removed from the inlet air. This has been mentioned in several of the environmental-type discussions above but is summarized here for completeness. Listed below are general recommendations. Gas turbine manufacturers may have their own set of guidelines for placing the gas turbine [29].

- When installing other combustion-type equipment, such as a diesel engine, near the gas turbine, the exhaust of the equipment should be directed away from the gas turbine inlet. This reduces the possibility of the exhaust gas entering the gas turbine inlet system. This exhaust can contain unburned hydrocarbons or corrosive gases.

- Cooling towers can be a major source of aerosol drift. Cooling towers are open to the atmosphere and, therefore, release aerosols into the air due to agitation from cross winds and the flow of the water down the tower. The water in the cooling tower also contains water treatment chemicals that could be detrimental to the gas turbine. The drift of aerosols from a cooling tower is confined within a few hundred feet. If possible, the gas turbine inlet should be positioned away from cooling towers and placed upstream of the prevailing wind direction to minimize the aerosol drift. CFD can be a useful tool to model how the wind will carry aerosols over to the gas turbine inlet. This will help the filter engineer to properly place the gas turbine to minimize cooling tower drift effects.

- Pressure relief valves are installed on many gas lines and equipment to protect the equipment in case of an over pressurization event. The vents to these relief devices should be directed away from the gas turbine inlet. Release of any hydrocarbon could result in high concentration ingestion at the filtration system. The filters at the inlet to the gas turbine do not remove gas phase contaminants.

- Piping connections on gas, fluid, or steam lines will generally leak after some time. The leaks at these connections can impact the filtration system. Piping should be routed away from the inlet in order to prevent this influence.

- Lube oil vents should be directed away from the inlet to prevent oil vapor ingestion.

- The exhaust of the gas turbine should be directed away from the inlet of the gas turbine. Carbon smoke and hydrocarbon fumes released at the exhaust could lead to accelerated fouling of the compressor blades.

- The gas turbine inlet system should not be directed toward or installed near any exhaust stacks. These exhaust stacks release chemical exhaust and unburned hydrocarbons, which can lead to compressor fouling and corrosion.

- Avoid placing the inlet near gravel or dirt roads. The dust thrown into the air from vehicle traffic and wind can be carried into the inlet of the gas turbine.(a)If the gas turbine is operated during construction activities, consider adding more robust filters to remove the excess dirt that will be ingested.

- Direct the inlet away from any open storage of coal, salt, or other grainy particles. The wind can carry the smaller grains from the storage area into the inlet of the gas turbine.

Site Evaluation

As discussed previously, there are several different types of environments where a gas turbine can operate. Also, there are many possible local, seasonal, and temporary contaminants that can be present. Therefore, each gas turbine installation site has a unique make-up of contaminants. When selecting the inlet filtration system, this make-up should be determined. This includes determining what contaminants and how much are present at the site. Once this information is known, the types of filters needed and filtration efficiency required can be established. Below is a list of items that should be considered when evaluating the site where the gas turbine will be installed [9, 29]:

- Environment where the gas turbine will be installed: Coastal, marine, offshore, desert, arctic, tropical, industrial area, rural countryside, or large city,
- Contaminants present in that environment,
- Local contaminants (mining operating, foundries, agricultural activities, inland salt lakes, etc.)
- Temporary contaminants (construction activity, dirt roads, etc.),
- Future emission sources (new industrial facility or residential development),
- Site layout (vents and exhaust, cooling tower drift, open storage of grainy particles, etc.),
- Weather patterns.

LIFE CYCLE COST ANALYSIS

When selecting a filtration system, the filter engineer is burdened with deciding the level of quality they want their system to achieve. This includes the efficiency of filtration, the particle size to be filtered, the amount of maintenance that will be needed to maintain the filtration system, what rate of degradation is acceptable for the gas turbine, the required availability and reliability of the gas turbine, what type of washing scheme will be used (online, offline, or a combination of both), and cost of the filtration system. The cost impact of each of the items mentioned can be quantified. A Life Cycle Cost (LCC) analysis provides a convenient means to compare different filtration system options quantitatively.

Life Cycle Cost Basics

This section covers the inputs that should be considered for the LCC analysis for a filtration system. It also provides methods to calculate the cost impact for each input. This type of analysis focuses on the overall or lifetime cost of a system. It is a tool that estimates the total cost to purchase, install, operate, maintain, and dispose of equipment. This analysis can assist in determining the best design options, which will minimize the overall cost of a system.

It is important to include initial cost in the analysis, but it is just as important to include operation and maintenances cost. The operating and maintenance cost over the life of a piece of equipment can have a more significant effect, especially if a poorly designed system is chosen. An LCC analysis can help to determine which system configuration can minimize lifetime costs. Some of the costs that are typically considered are shown below. Examples of how this would apply to filtration systems are provided in parentheses:

- Initial cost (filters, filtration system, spares filters, instrumentation),
- Installing and commissioning costs (labor, cost of installation equipment (such as cranes), shipping costs),
- Energy costs (pulse system for self-cleaning filters),
- Operating costs (labor, inspections),
- Maintenance (replacing filters, repairing system, labor for maintenance),
- Downtime (replace filters, complete offline water washes, anything outside of normal shutdowns for other maintenance),
- Gas turbine effects (degradation, performance loss),
- Decommissioning and disposal (disposal of filters).

In an LCC analysis, estimates are provided for each cost component of the system. An inflation rate can be applied to the costs which will occur later in the life of a system (such as 10 years from the installed date). Once these costs are established, they are brought back to present value using (2). The Net Present Value (NPV) term represents the value of the cost in present terms. A is the value of the cost in the year it occurs. The term i is the discount rate and n is the year the cost occurs in. If there is a price increase (inflation) or decrease, then this can be accounted for by using (3). The term e is the increase or decrease in price:

$$NPV = A\,(1+i)^n \tag{2}$$

$$NPV = A\,(1+(i-e))^n \tag{3}$$

Projected costs over the lifetime of the system cannot be combined directly when calculating the LCC, because the funds spent at different times have different values to the investor. The discount rate, i, is used to bring the costs to present terms, where they can be directly added together, and is defined as the rate of return that is used to compare expenditures at different points in times. For example, the investor would be equally satisfied to have one amount received earlier and the other amount received later.

If a cost occurs yearly, the NPV of the total recurring costs can be calculated with (4). If inflation or price escalation is considered in the analysis, the NPV of the total recurring cost can be calculated with (5):

$$NPV = A/i(1 - \lfloor 1+i \rfloor^n)$$
(4)

$$NPV = A(1+e/1-e)(1 - \left[1+e/1+i\right]^n)$$
(5)

The NPVs must be determined for each cost. Then the cost will be added together to obtain the total NPV or LCC cost [30].

Considerations for an Inlet Filtration System

In an LCC analysis for a gas turbine inlet filtration, there are six main parameters: purchase price/initial cost, maintenance cost, availability/reliability of the gas turbine, gas turbine degradation and compressor washing, pressure loss, and failures of the filtration system cr gas turbine due to inlet air quality [29].

Purchase Price/Initial Cost

The purchase price occurs in the first year of the LCC analysis. It is the cost to purchase and install the inlet filtration system. An estimate for this value can be obtained from the filter vendor or gas turbine manufacturer.

Maintenance Cost

The maintenance cost includes the cost of filter replacement and disposal and any maintenance to auxiliary systems for the inlet filtration system. It is a recurring cost that should be included in each year that the cost is acquired. This cost can be calculated based on estimated filter change out frequencies, cost of filters from vendors, labor cost for maintenance, and cost of downtime to replace filters.

Availability/Reliability of Gas Turbine

The availability/reliability of a gas turbine impacts the cost due to the lost production as a result of the nonavailability of the gas turbine. Filter exchanges requiring the shutdown of the engine, as well as on-crank water washing negatively, impacts the availability of the engine. On the other hand, if the engine is not used 100 percent of the time, for example, because it is a standby or peaking unit, the cost of degradation has to be adjusted accordingly.

Gas Turbine Degradation and Compressor Washing

Gas turbine degradation is perhaps the most important cost in the analysis. This is often the cost which drives the analysis to favor one filtration system option over another. The cost of gas turbine degradation is calculated based on the reduced power output and increased heat rate due to inlet air quality. The rate of degradation due to inlet air quality is difficult to calculate and is best found from past operating history. There are several degradation models discussed in the literature, which can provide estimates of the expected degradation rate. A few examples are the models presented by Zaba and Lombardi [31], Kurz and Brun [32], and Meher-Homji et al. [33].

Once the degradation rate is calculated the lost profit due to reduced gas turbine output can be calculated. If the gas turbine is operating at full load, then it is expected that the fuel cost will decrease due to the lower power output. For part load operations, it is expected that the fuel cost will go up since the engine will be operated at the desired power output. The change in fuel cost should be calculated based on the change in heat rate and operational philosophy and be included in the analysis. This cost should be included in each year of the analysis.

Compressor washing is often performed in a gas turbine in order to minimize the effects fouling on the performance of the gas turbine. The use of compressor washing may reduce the rate of degradation in the gas turbine. However, the most effective type of washing is on-crank washing, which requires that the engine is shut down. This results in a lower availability of the engine, and, associated with this, may cause the cost of lost production.

Pressure Loss

The pressure loss across the inlet filtration system can also have a significant effect on the cost of the inlet filtration system. An increase in the pressure loss across the filtration system leads to reduced power output from the gas turbine and an increased heat rate. The cost of these effects should be included yearly in the LCC analysis.

Failure/Event Cost

The last cost is any cost associated with a failure or event that occurs due to the inlet filtration system or inlet air quality. This could be a failure of a filter material, which requires shutdown for replacement or a failure of a gas turbine blade which occurred due to corrosion from poor inlet air quality. These costs are often included based on past experience with the gas turbine or other filtration systems.

CONCLUSIONS

In summary, the selection and operation of an inlet filtration system is highly dependent on the environment where the gas turbine is operating. The contaminant present in the ambient air will dictate the type filters that are used. It is important to quantify what type and size of contaminants are present in order to correctly select the filters to be used. Temporary and seasonal variations must also be considered for the inlet filtration system. A life cycle cost analysis provides a convenient method to quantify and compare various filtration system options such that the optimal system can be selected.

REFERENCES

1.   R. Kurz and K. Brun, "Gas turbine tutorial—maintenance and operating practices effects on degradation and life," in Proceedings of the 36th Turbomachinery Symposium, 2007.

2.   P. T. McGuigan, "Salt in the marine environment and the creation of a standard input for gas turbine air intake filtration systems," in Proceedings of the ASME Turbo Expo Power for Land, Sea, and Air, Vienna, Austria, 2004, GT2004-53113.

3.   S. Howes, "Selecting gas-turbine inlet air systems for new, retrofit applications," Combined Cycle Journal, Second Quarter. 2004.

4.   E. Syverud, O. Brekke, and L. E. Bakken, "Axial compressor deterioration caused by saltwater ingestion," Journal of Turbomachinery, vol. 129, no. 1, pp. 119–126, 2007. View at Publisher · View at Google Scholar · View at Scopus

5.   T. Z. Baden, "Losses in gas turbines due to deposits on the blading," Brown Boveri Review, vol. 67, no. 12, pp. 715–722, 1980. View at Scopus

6.   HEPA Filtration Facts, Donaldson Filtration Solutions, 2009.

7.   Principles of Air Filtration, Mueller Environmental Design, 2009.

8.   M. Owens, Engineering Bulletin—Compressor Fouling Benhmark, AAF International, 2009.

9.   R. L. Loud and A. A. Slaterpryce, "Gas Turbine Inlet Air Treatment," GE Power Generation, GER-3419A, 1991.

10. ASHRAE 52.2, Method of Testing General Ventilation Air-Cleaning Devices for Removal Efficiency by Particle Size, American Society of Heating, Refrigeration and Air-Conditioning Engineers, 2007.

11. DIN EN 779, Particulate Air Filters for General Ventilation—Determination of the Filtration Performance, European Committee for Standardization, 2002.

12. DIN EN–1 1822, High Efficiency Air Filters—Part 1: Classification, Performance Testing, Marking, European Committee for Standardization, 2009.

13. DIN EN–2 1822, High Efficiency Air Filters—Part 2: Aerosol Production, Measuring Equipment, Particle Counting Statistics, European Committee for Standardization, 2009.

14. DIN EN–3 1822, High Efficiency Air Filters—Part 3: Testing Flat Sheet Filter Media, European Committee for Standardization, 2009.

15. DIN EN–4 1822, High Efficiency Air Filters—Part 4: Determining Leakage of Filter Element (Scan Method), European Committee for Standardization, 2009.

16. DIN EN–5 1822, High Efficiency Air Filters—Part 5: Determining The Efficiency of Filter Element, European Committee for Standardization, 2009.

17. R. Kurz and K. Brun, "Fouling mechanisms in axial compressors," in Proceedings of the ASME Turbo Expo, Power for Land, Sea, and Air, Vancouver, Canada, 2011, GT2011-45012.

18. Gas Turbine World 2009 GTW Handbook, vol. 27, 2009.

19. T. J. Retka and G. S. Wylie, "Field experience with pulse-jet self-cleaning air filtration on gas turbines in an arctic environment," Journal of Engineering for Gas Turbines and Power, vol. 109, no. 1, pp. 79–84, 1987. View at Scopus

20. A. Klink and T. Schroth, "New solutions for improved intake air filtration of gas turbines and turbocompressors," in Proceedings of the ASME Turbo Expo Power for Land, Sea, and Air, Birmingham, UK, 1996.

21. R. K. Mudge and S. D. Hiner, "Gas turbine intake systems—high velocity filtration for marine gas turbine installation," in Proceedings of the ASME Turbo Expo, Power for Land, Sea, and Air, New Orleans, La, USA, 2001, 2001-GT-0584.

22. A. D. Oswald and S. D. Hiner, "More efficient applications for naval gas turbines—addressing the mismatch between available technology and the requirements of modern naval gas turbine inlets," inProceedings of the ASME Turbo Expo Power for Land, Sea and Air, Barcelona, Spain, 2006, GT2006-90305.

23. J. P. Stalder and J. Sire, "Salt percolation through gas turbine air filtration systems and its contribution to total contaminant level," in Proceedings of the International Joint Power Generation Conference, pp. 445–456, New Orleans, La, USA, June 2001, JPGC2001/PWR-19148. View at Scopus

24. R. G. Neaman and A. W. Anderson, "Development and operating experience of automatic pulse-jet self-cleaning air filters for combustion gas turbines," in Proceedings of the Gas Turbine Conference and Products Show, New Orleans, La, USA, 1980, 80-GT-83.

25. O. Brekke, L. Bakken, and E. Syverud, "Filtration of gas turbine intake air in offshore installations: the gap between test standards and actual operating conditions," in Proceedings of the ASME Turbo Expo, Power for Land, Sea, and Air, Orlando, Fla, USA, 2009, GT2009-59202.

26. D. G. Hill, "Gas turbine intake systems in unusual environments," in Proceedings of the Gas Turbine Conference and Products Show, Washington, DC, USA, 1973, 73-GT-38.

27. C. Brake, "Identifying areas prone to dusty winds for gas turbine inlet specification," in Proceedings of the Turbo Expo, Power for Land, Sea, and Air, Montreal, Canada, 2007, GT2007-27820.

28. R. E. Cleaver, "Gas turbine filtration in tropical environments," in Proceedings of the Turbomachinery Maintenance Congress, 1990.

29. M. Wilcox, R. Baldwin, A. Garcia-Hernandez, and K. Brun, "Guideline for gas turbine inlet air filtration systems," in Proceedings of the Gas Machinery Research Council, 2010.

30. Europump and Hydraulic Institute, Pump Life Cycle Costs: A Guide to LCC Analysis for Pumping Systems, 1st edition, 2001.

31. T. Zaba and P. J. Lombardi, "Experience in the operation of air filters in gas turbine installation," Brown Boveri Review, vol. 72, no. 4, pp. 165–171, 1985. View at Scopus

32. R. Kurz and K. Brun, "Degradation in gas turbine systems," Journal of Engineering for Gas Turbines and Power, vol. 123, no. 1, pp. 70–77, 2001. View at Publisher · View at Google Scholar

33. C. B. Meher-Homji, M. Chaker, and A. F. Bromley, "The fouling of axial flow compressors—causes, effects, susceptibility and sensitivity," in Proceedings of the ASME Turbo ExpoPower for Land, Sea, and Air, Orlando, Fla, USA, 2009, GT2009-59239.

34. Offshore Filter Systems, Camfil Farr Brochure, 2009.

35. Cam GT for Turbomachinery, Camfil Farr Product Sheet, 2009.

36. Industrial Filtration, Burgess-Manning Bulletin BM-2-301A, 2009.

Control System Design for a Gas Turbine Engine Using Evolutionary Computing for Multidisciplinary Optimization

Valceres V. R. e Silva[I], Wael Khatib[II],
and Peter J. Fleming[II]

[I]Universidade Federal de São João del Rei - Praça Frei Orlando 170 -
36307 352 - São João del Rei - MG
[II]The University of Sheffield Mappin Street, S1 3JD - Sheffield - UK

ABSTRACT

Multidisciplinary optimization (MDO) is concerned with complex systems exhibiting challenges in terms of organization and scale. Thus, it is well suited to be applied to complex multivariable control design. Collaborative optimization is one approach for dealing with complex multidisciplinary optimization problems. Three MDO architectures, including collaborative optimization, are applied to control system design for a gas turbine engine, in order to improve the design search process by exploring possible solutions with parallel, but independent search strands. The optimization is carried out through a multiobjective genetic algorithm framework.

INTRODUCTION

There is a significant body of research devoted to the study of design and optimization of a number of interacting or coupled systems. Most of this research tends to be related to aerostructural design and is called multidisciplinary optimization. The design of an airplane requires the bringing together of resources representing structures, metallurgy, aerodynamics, performance, control and other disciplines in order to produce an optimal design. The main challenges faced in MDO design problems are computational cost and organizational complexity (Sobieszczanski-Sobieski and Haftka, 1996). The complexity of design optimization depends on the complexity of the pertinent disciplines, the size of the problem, and the nature of the objectives and constraints. Comparing with an aggregation of many disciplines, the problem grows very much in complexity, if there is more than one discipline controlling the same design variables for a particular objective. This is mainly due to the effect of coupling between the variables. Organizational complexity is due to the fact that the various disciplines traditionally reflect different analysis methods, schools of thought, software and hardware platforms, standards, etc. The organizational challenge in MDO is for an efficient exchange of data, systems integration and other aspects of communication.

Evolutionary computing refers to computer-based problem solving systems that use evolutionary algorithms (EAs). EAs generally use computational models that exploit mechanisms based on the neo-Darwinian theory of evolution. The main techniques used in EAs include: genetic algorithms (GAs), evolutionary programming (EP), evolution strategies (ES) and genetic programming (GP). EAs are amenable to parallelization and can help reduce the computational cost. These algorithms are stochastic in nature and can usually start an optimization process without much a priori knowledge. No derivative information is required as in the traditional gradient based methods and this helps EAs deal with difficult search spaces characterized by multimodal disjoint feasible areas.

Most design problems are multiple objectives in nature, including MDO problems. These objectives are often conflicting or competing. The concept of Pareto optimality is a powerful method for dealing with multiple objectives. Using this approach, the designer is no longer searching for a single optimum, rather a compromise satisfying the various objectives. and constraints. The collection of compromise solutions is referred to as the non-dominated set. Within this set, attempted improvement in one objective will result in degradation in one or more of the others. EAs are amenable to multiobjective optimization (MO). This is because an EA works on a population of solutions instead of the traditional single point search. The search with this population can help achieve a faster and more comprehensive mapping of the trade-off hyper surface. An overview of the application of EAs to MDO can be found in Khatib and Fleming (1997).

THE GAS TURBINE ENGINE

Gas turbine engines (GTE) are highly nonlinear plants that have multiple inputs and outputs. The operating conditions span extremes of temperature, pressure and load conditions. The engine performance requirements cover a wide flight envelope that includes a continuum of set points of altitude and speed in terms of the Mach number. These requirements add to the complexity of designing suitable controllers that can achieve high performance levels while maintaining stability and safe operation with minimum overall cost. H_∞ and PI (proportional and integral) controllers have been designed for the GTE using simplified models obtained through the response surface variable complexity modelling technique (Silva et al., 2007). It was also obtained inprovements on this engine's performance by reducing fuel consumption, increasing thrust in dash missions and minimizing turbine blade temperature (Silva et al., 2005). In this work, multidisciplinary collaborative optimization structures are used to split the PI controller design problem in three and thus, optimization is carried out.

The engine model has three inputs: fuel flow (wFE), exhaust nozzle area (A8) and inlet guide vane angle (IGV). Sensors provided from engine outputs include: high pressure spool speed (NH), low pressure spool speed (NL), engine pressure ratio (EPR) and fan pressure ratio. These measurements can be used to provide various pairings of input-output for closed-loop control. Important engine variables such as thrust (XGN) and surge margin (LPSM) cannot be measured directly. Such variables are controlled implicitly through other related measurable values such as pressure ratios and bypass duct Mach number (DPUP). Silva and Fleming (2002) used a non-linear model for control configuration and PI (proportional and integral) controller design, and also for a H_y controller design. Using findings from these studies, a closed-loop control strategy is chosen for parametric optimization of a designed PI controller structure for a particular operating point to demonstrate the use of MDO architectures for control design (Fig. 1).

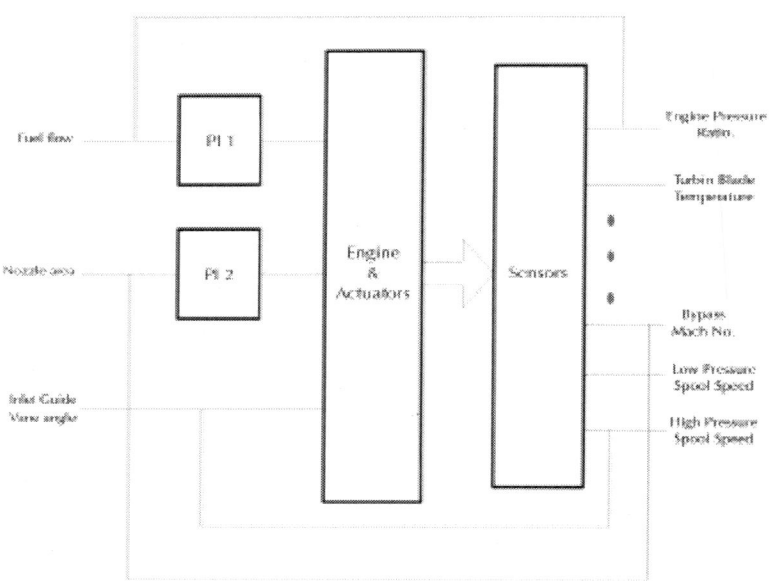

Figure 1: PI controller structure for the spey GTE.

Two PI controllers are implemented for controlling XGN and LPSM. Because these two variables cannot be measured directly, they are controlled implicitly through EPR and the DPUP respectively. The third input, inlet guide vane angle (IGV), is gain scheduled against the measured output values of the NH. The two PI controllers supplied with the Rolls-Royce SIMULINK model of this engine, use the structure proposed by Åström and Häagglund (1995), and it is shown in Figure 2.

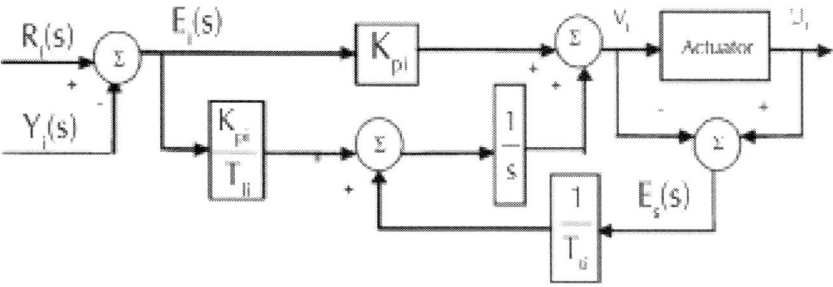

Figure 2: PI controller with anti-windup.

When there is no actuator saturation, the anti-windup feedback signal $E_i(s)$ is zero The expression of the output U_i of the i-th controller is given by equation 1.

$$U_i(s) = K_{pi}E_i(s) + \frac{K_{pi}}{T_{Ii}s}E_i(s)$$

(1)

where, $E_i = R_i - Y_i$, is the error signal for loop i, R_i is the set point signal for loop i, and i = 1, 2. With reference toFigure 2, K_{pi} and T_{li} denote the parameters of the i-th PI controller. The parameter T_{ti} is known as the tracking time constant, and controls the effect of the integral anti-windup mechanism. This structure has been applied to design optimized PI controllers with genetic algorithm for industrial plants (Ghaffari, 2007).

The design is expected to satisfy the following objectives and constraints:

- XGN > 48.6 kN (performance requirement).
- TBT < 1713 °C (physical limit).
- LPSM > 10% (stability limit).
- XGN rise and settling time for speed of response performance.
- steady state error limits for NH and NL.
- steady state error limit for exhaust nozzle area (A8).

- limits on overshoot/undershoot for transient regions of NH and NL.
- other engine limits for spool speeds and exhaust nozzle area.

A successful design of an engine controller can be achieved as part of, say, an overall MDO design of an aircraft. Such designs are still not attempted on a full scale in industry due to the limitations of available resources for MDO. However, MDO techniques can be applied to the design process for the controller per se. To demonstrate this, we present an example for designing a PI control strategy for a GTE that comprises two PI controllers. A nonlinear SIMULINK model for a Spey GTE is used to design and evaluate controller performance.

MULTIOBJECTIVE GENETIC ALGORITHM

Multiobjective optimization and decision making refers mainly to simultaneous optimization in order to achieve optimal trade-off solutions satisfying various objectives. These objectives tend to be conflicting or competing. There is not usually one unique solution but rather a family of compromise solutions that need to be analyzed by a decision maker.

The multiobjective genetic algorithm (MOGA) combines the characteristics of a powerful evolutionary optimization strategy, the genetic algorithm with the concept of Pareto optimality (Giannakoglou, 2002) to produce solutions illustrative of a problem's trade-off set. A MOGA evolves a population of solution estimates thereby conferring an immediate benefit over conventional multiobjective optimization methods.

Mathematically, the multiobjective optimization (MO) problem is to find a vector of design variables x, that is within the feasible region in the universe $\hat{A}$, to minimize (or maximize) a vector of objective functions F(x). Some or all of the component functions can be non-linear. Most practical problems are also bounded by a vector of constraints g(x). Multiobjective optimization can be expressed as follows:

$$\text{Minimize: } F(x) = \{f_1(x), f_2(x), \ldots, f_n(x)\}$$
$$\text{subject to: } g(x) \leq 0$$

(2)

where g(x) is the constraint vector and $f_i(x)$ is the i-th objective function.

The set of trade-off solutions that express the best performance in all of the objectives is known as the Pareto or the non-dominated set.

The concept of Pareto-optimality constitutes by itself the origin of research in multiobjective optimization. In a multiobjective minimization problem, a feasible vector x* ∈ X is Pareto-optimal if and only if there is no feasible vector x ∈ X such that for all i ∈ {1, 2, ..., n}

$$f_i(x^*) \leq f_i(x)$$

(3)

and for at least one i ∈ {1, 2, ..., n}

$$f_i(x^*) < f_i(x)$$

(4)

The decision making process picks the best solution from the non-dominated set (Pareto-optimal) according to some preference information.

Most real-world optimization problems are multi-modal. There often exist several criteria to be considered by the designer. The compromise of better performance for all of them has to be achieved. Fonseca and Fleming (1995) use the ranking approach for assigning fitness to each individual in the population. They define the individual's rank simply as the number of members of a population in a generation that dominate it. Thus, non-dominated individuals are assigned rank zero, while the lowest possible rank in any generation of population is rank n-1, where n is the number of individuals in the population. The fitness is then assigned to each individual by interpolating from the best to the worst, according to some function, that can be linear, exponential or other type.

Any attempted improvement for a member of this set in one of the objectives will result in deterioration in performance in one or more of the other objectives.

The work described here employs a genetic algorithm with an implementation of multiobjective optimization as proposed by Fonseca and Fleming (1993).

MDO ARCHITECTURES

Balling and Sobieszczanski-Sobieski (1996) introduce a consistent method for classifying the various approaches for formulating MDO problems using compact and consistent notation. For practical purposes, the various approaches can be grouped into three main categories:

- Single-level approaches
- Collaborative optimization
- Concurrent sub-space optimization

A detailed discussion of these methods can be found in Khatib and Fleming (1998). Single-level approaches address a design problem as one whole unit. Concurrent sub-space optimization relies on recursive iteration loops and can be less efficient. The collaborative optimization (CO) approach works through decomposition of complex large-scale problems into smaller sub-problems or elements. Each element proceeds with its own optimization using separate decision variables. The outputs of the various disciplines can be pooled into a shared resource area to be redistributed or observed by a system designer or optimizer to ensure a viable overall design. The decomposition boundaries depend on the physical organization of the problem, the available resources and/or the mathematical limits. This gives rise to soft and hard boundaries. The evolutionary collaborative architecture proposed by Khatib and Fleming (1998) allows the various disciplines to progress simultaneously in search of optimal designs. This approach draws on a seemingly good match between the two elements of CO and EAs. They share amenability to parallelization and this is also reflected in existing practices in industry.

MOGA-PI DESIGN ARCHITECTURE

For single level optimization of a MIMO (Multi Input Multi Output) system, in terms of the underlying GA engine in the MOGA (multiobjective genetic algorithm) framework, the two main operators that guide the optimization process are selection and crossover. The selection pressure which favors the fitter individuals is based on how well each of the individuals performs in terms of Pareto optimality. Grouping the proportional and integral terms for both controllers together for each individual suggests that an integral term, say, might be promoted in the population based on influence of a proportional term in the same individual. This form of nepotism can be inefficient in computing terms and might even lead to less optimal solutions.

Three different frameworks using MOGA are applied, employing one single-level and two collaborative design approaches. The single-level MOGA addresses the two controllers and all the objectives simultaneously. Each individual in the MOGA population is a chromosome made up of the four controller parameters in a Gray-coded binary representation. The two CO implementations are different in the way the design problem is decomposed.

CO I has two optimization sub-problems: one dealing with the proportional gains in both controllers, and the other dealing with the integral gains. The objectives for each of these sub-problems reflect potential scope of influence. The proportional gains are designed to optimize all the objectives and constraints of the problem simultaneously, while the integral gains are considered for optimization of XGN and TBT dynamics only. The optimization process for P1 and P2 can be described as follows:

$$
\begin{aligned}
\text{Maximise:} \quad & \text{LPSM} \\
\text{Minimize:} \quad & \text{NH and NL steady-state errors} \\
& \text{A8 steady-state error} \\
& \text{NH and NL overshoot/undershoot.} \\
\text{Satisfy:} \quad & \text{XGN} \geq 48.6 \text{ kN} \\
& \text{TBT} \leq 1713\,^{\circ}\text{C} \\
& \text{XGN rise time} \leq 1.0 \text{ s} \\
& \text{XGN settling time} \leq 1.4 \text{ s} \\
& \text{NH} \leq 102\% \\
& -8^{\circ} \leq \text{IGV} \leq 32^{\circ} \\
& 0.25 \text{ m}^2 \leq \text{A8} \leq 0.34 \text{ m}^2
\end{aligned}
\tag{5}
$$

The optimization process fo I1 and I2 has no soft objectives. It is a constraint satisfaction problem in which MOGA is used to satisfy the following constraints:

$$TBT \leq 1713°C$$
$$XGN \text{ rise time} \leq 1.0 \text{ s}$$
$$XGN \text{ settling time} \leq 1.4 \text{ s}$$
$$NL \leq 102\%$$
$$-8° \leq IGV \leq 32°$$
$$0.25 \text{ m}^2 \leq A8 \leq 0.34 \text{ m}^2 \tag{6}$$

CO II designed for this problem depicts the boundaries by having each of two sub-problems dealing with one of the two PI controllers. The pertinent objectives also reflect the level of influence: PI1 design optimizes XGN, NH and NL dynamics (objectives d and g), steady state errors for NH and NL (objectives e) whilst minimizing TBT (objective b). PI2 design optimizes the A8 variations (objective f) while maintaining LPSM and TBT within safe limits (objectives b and c). It can be noticed that TBT is common to both controllers. This suggests coupling in the system and can help the two design sub-problems achieve a satisfactory overall results. The optimization process for PI1 can be described as follows:

Minimize: NH and NL steady-state errors
           NH and NL overshoot/undershoot.
Satisfy: $XGN \geq 48.6$ kN
           $TBT \leq 1713°C$
           XGN rise time ≤ 1.0 s
           XGN settling time ≤ 1.4 s
           $NL \leq 102\%$
           $-8° \leq IGV \leq 32°$
           $0.25 \text{ m}^2 \leq A8 \leq 0.34 \text{ m}^2$

$$\tag{7}$$

The optimization process for PI2 can be described as follows:

Maximise: LPSM
Minimize: A8 steady-state error
Satisfy: $TBT \leq 1713°C$
           $NL \leq 102\%$
           $-8° \leq IGV \leq 32°$
           $0.25 \text{ m}^2 \leq A8 \leq 0.34 \text{ m}^2$

$$\tag{8}$$

The work described here employs a GA with an implementation of MO as proposed by Fonseca and Fleming (1993). This multiobjective GA (MOGA) presents a simple but powerful tool for design optimization were preferences can be articulated progressively by the decision maker as more insight into the problem in hand is gained.

IMPLEMENTATION

The three different frameworks described in the previous section are used for parametric optimization of a PI controller structure. The MOGA generates two sets of PI controller gains for the wFE and A8 loops. The MOGA parameters for crossover and mutation are 90% and 5% respectively.

Because of the parallel nature of MOGA, the two sub-problems in CO I and CO II run concurrently. Each problem has its own population of solutions. Each population experiences selection and crossover driven by its own pertinent objectives. The two populations do not share or individuals through migration. To achieve a consistent overall design, each population presents its proposed solutions to a shared data pool which performs the engine simulations (Figure 3). Results of the simulations are fed back into the pool and each separate MOGA scans for its relevant outputs to drive its own GA operators.

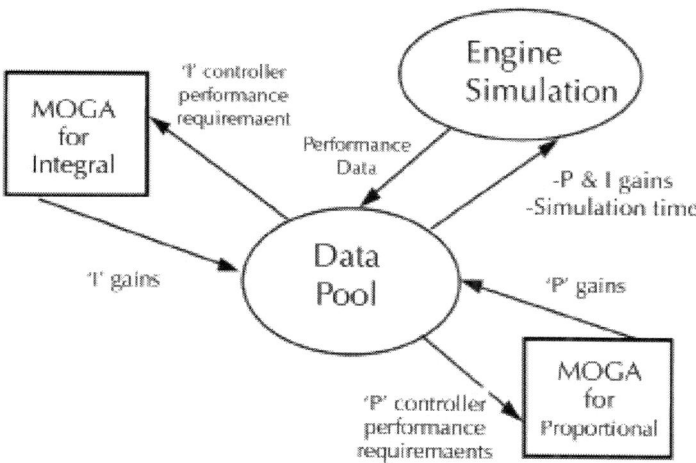

Figure 3: MOGA CO I architecture.

The control response is measured after a step input in thrust demand, corresponding to changes from 85% to 90% of high pressure spool speed. Using two populations of solutions simply increases the requirements to compute the MOGA operators for ranking, selection, crossover and mutation. These costs are negligible when compared to the cost of engine simulations. The number of simulations required is the same as for the single-level optimization case. Each of the three MOGAs employs 40 individuals and is evolved for 50 generations.

RESULTS

The multidisciplinary optimization and multiobjective optimization produce a family of non-dominated solutions, for each of the three aproaches: 21, 12 and 12 controller sets for the single-level, CO I and CO II respectively.Figures 4 - 5 present XGN step responses for changes from 35% to 87% of thrust demand, for the obtained controllers of the three design architectures. The responses are normalized for the desired point. Most of these controllers exhibit similar performance characteristics for the three structures. For the CO II approach, the responses are slower than for the other two cases. Although for the single-level, the responses seem to be the fastest, it can also be seen that some responses for the CO I structure outperformed the single level, even considering this criterium.

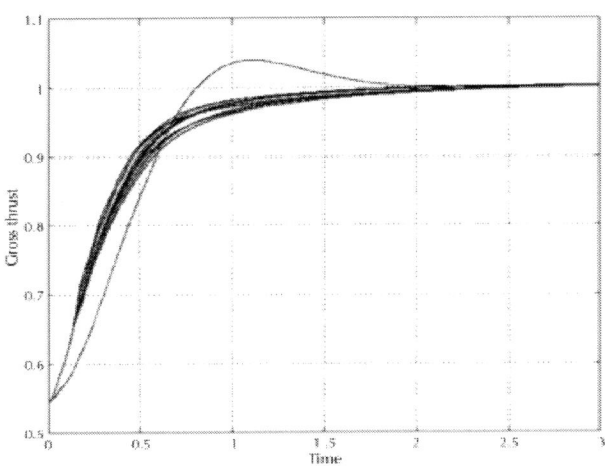

Figure 4: Thrust responses for single-level approach.

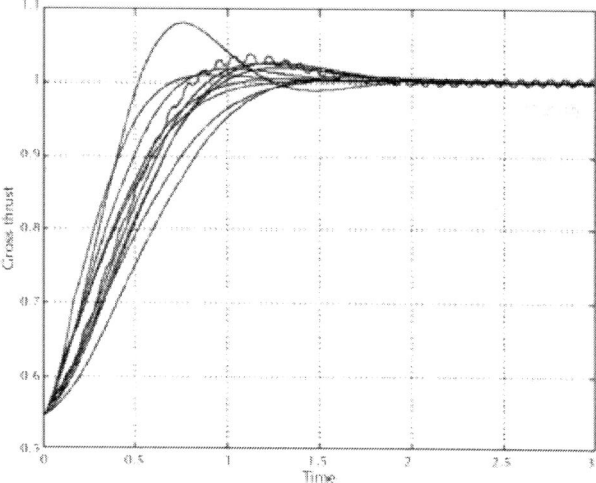

Figure 5: Thrust responses for CO I approach.

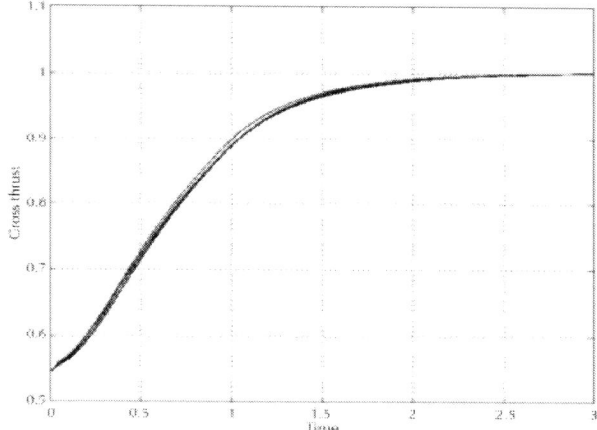

Figure 6: Thrust responses for CO II approach.

To study the control performance in more detail, further filtering is applied to choose one controller set for analysis from each approach. The full thermodynamic model is evaluated for all the controllers in each set. The objectives representing the steady-state errors are all catered for adequately by all solutions and are ignored. The solutions are now further ranked but only using a subset of the objectives a - d only, to choose the best controller using multiobjective ranking. All the performance graphs (Fig. 7 - 10) refer to the final selected controller set for each case. They indicate that the CO I controller outperforms the other two. CO II perfomance indicates slower, though stable, responses when compared to the other two. The superiority of CO I was found to hold for all the other nondominated controllers in each case.

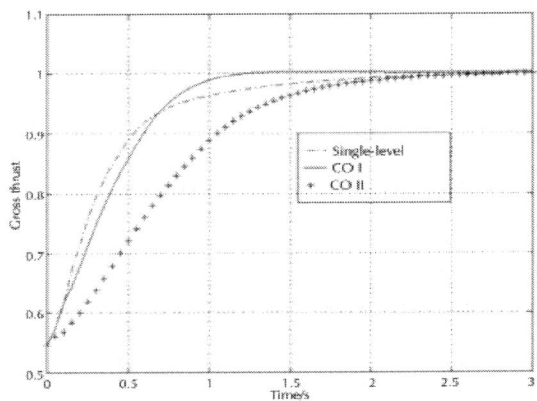

Figure 7: Thrust responses for the three design approaches.

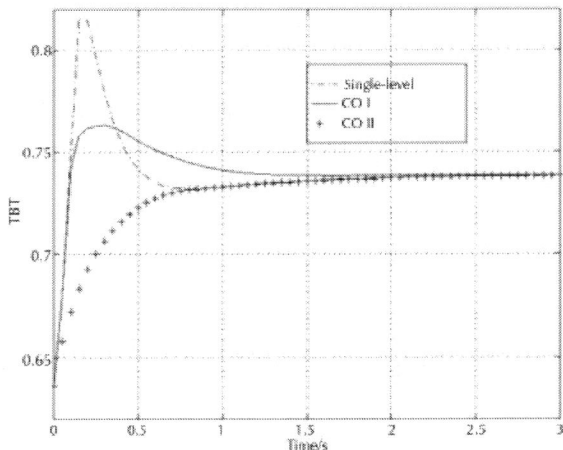

Figure 8: TBT responses for the three design approaches.

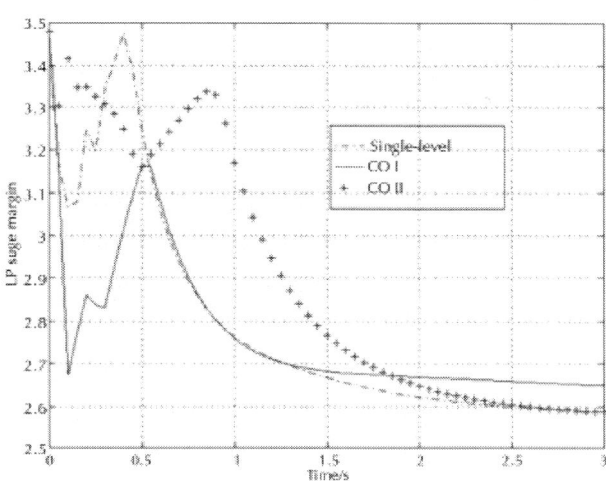

Figure 9: LPSM responses for the three design approaches.

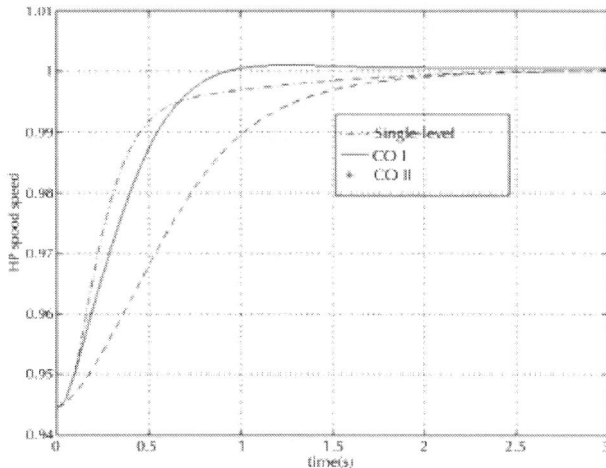

Figure 10: NH responses for the three design approaches.

In all the response graphs, CO II appears to indicate the slowest performance indicating overdamping, but the responses are still within limits. The thrust response indicates that the single-level controller achieves slightly faster response. The transient behavior however, appears to indicate that CO I gives the most satisfactory results overall. LPSM and TBT overshoot are worst for the single-level case.

Thus, decomposing the problem into the two different types of controller (P and I) as in CO I gives better results than spliting the actual controllers along the structures corresponding to different input-output combinations as in CO II. This decomposition also outperforms the classical single-level all in one approach. The engine is one whole unit made up of various interacting sub-units. The single level approach can allow some perfomance dynamics, pertinent only to a subset of the controller parameters, to influence other controller parameters. If this is the case, it will lead to inefficiencies. CO I appears to offer a better decomposition scheme than CO II. This allows the relevant perfomance metrics to be allocated to each optimization strand in a more efficient and appropriate way.

In order to calculate the computational cost of MOGA-PI approach, it is useful to consider relative computing costs (in seconds of CPU time) of the design stages as follows:

- Model evaluations

  PI stable model evaluation: $a = 27.8383$ s/evaluation

  PI unstable model evaluation: $b = 11.4950$ s/evaluation

- Optimization costs: there are five operators used in the MOGA: selection, crossover, mutation, multiobjective ranking and fitness assignment. The average cost of all of these operators is: $c = 0.006167$ s/individual

The optimization was carried out over 50 generations of 40 individuals each. Thus, the computational cost is: $c = 50*40*(a + b)/2 + 50*40*c = 39345.624$ seconds

The evaluation of the models is the main cost of the overall optimization process.

CONCLUSIONS

Multidisciplinary optimization is used to tackle some difficult design problems involving interacting systems with strong coupling, and is traditionally based in the aircraft industry.

Complex systems tend to have strong coupling between the various elements, creating complex optimization scenarios ir control design problems, and are usually multidisciplinary in nature. MDO challenges are envisaged to be present in complex control scenarios such as multivariable control.

This is the case for the optimization of the PI controllers for a GTE. The design methodology using collaborative optimization together with MOGA decomposes the problem along convenient and efficient boundaries. This allows for improvements in terms of computational cost, problem set-up and efficiency issues, and can also lead to improvements in the performance of the final designs.

Some improvements are possible using a collaborative optimization framework for PI controllers design. These improvements are reflected in the actual control performance in the engine response. The compromise solutions showed the trade-offs between the variables and can help the designer understands the interactions between the sub-systems. The final overall designs can also offer improvements over traditional single-level all-in-one approaches.

REFERENCES

1. Åström, K.J. and Häagglund, T. (1995), "PID controllers: Theory design and tuning", (2nd Edition), Instrument Society of America.

2. Balling, R.J. and Sobieszczanski-Sobieski, J., (1996). "Optimization of Coupled Systems: A Critical Overview of Approaches". AIAA Journal, 34(1), pp. 6-17.

3. Chipperfield, A.J. and Fleming, P.J., (1996). "Systems integration using evolutionary algorithms". International Conference on Control'96 - IEE Conference Publication n° 427, 1, pp. 705-710.

4. Fonseca, C.M. and Fleming, P.J., (1993). "Genetic Algorithms for Multiobjective Optimization: Formulation, Discussion and Generalisation". In Genetic Algorithms: Proceedings of the Fifth International Conference (S. Forrest, ed.), San Mateo, CA: Morgan Kaufmann.

5. Fonseca, C.M. and Fleming, P.J., (1995). "An overview of evolutionary algorithms in multiobjective optimization". Evolutionary Computing. Vol. 3, N°. 1, pp. 1-16.

6. Ghaffari, A., Mehrabian, A.R. and Mohammad-Zaheri, M., (2007). "Identification and control of power plant de-superheater using soft computing techniques". Engineering Applications of Artificial Intelligence, Vol. 20, Issue 2, pp. 273-287.

7. Giannakoglou, K.C., (2002), "Design of optimal aerodynamic shapes using stochastic optimization methods and computational intelligence", Progress in Aerospace Sciences, Vol. 38-1, pp. 43-76.

8. Khatib W. and Fleming, P.J., (1998). "Evolutionary Computing Applied to MDO Test Problems". 7th AIAA/USAF/NASA/ISSMO Symposium on Multidisciplinary Analysis and Optimzation, St. Louis, MO.

9. Khatib W. and Fleming, P.J., (1997). "Evolutionary computing for multidisciplinary optimization". Proc 2nd IEE/IEEE International Conference on Genetic Algorithms in Engineering Systems: Innovations and Applications GALESIA'97, Glasgow, pp 7-12.

10. Silva, V.V.R., Fleming, P.J., Sugimoto, J, and Yokoyama, R., (2007). "Multiobjective optimization using variable complexity modelling for control system design". Applied Soft Computing, In Press, Corrected Proof, Available online 18 March 2007.

11. Silva, V.V.R., Khatib, W. and Fleming, P.J., "Performance optimization of gas turbine engine". Engineering Applications of Artificial Intelligence, Vol. 18, Issue 5, August 2005, pp. 575-583.

12. Silva, V.V.R. and Fleming, P.J., (2002). "Control configuration design using evolutionary computing". Proceedings of XV Triennial IFAC World Congress. Barcelona, Spain.

13. Sobieszczanski-Sobieski, J. and Haftka, R.T., (1996). "Multidisciplinary Aerospace Design Optimization: Survey of Recent Developments". AIAA Paper 96-0711, Reno, NV.

Chapter 4

The Thermal Stability of Eyjafjallajökull Ash versus Turbine Ingestion Test Sands

Ulrich Kueppers[1], Corrado Cimarelli[1], Kai-Uwe Hess[1],
Jacopo Taddeucc[i2], Fabian B Wadsworth[1],
and Donald B Dingwell[1]

[1]Earth & Environmental Sciences, Ludwig-Maximilians-Universität München, Theresienstr. 41, 80333 Munich, Germany

[2]Istituto Nazionale di Geofisica e Vulcanologia, Via di Vigna Murata 605, 00143 Roma, Italy

ABSTRACT

The 2010 eruption of Eyjafjallajökull (Iceland) and the 2011 eruptions of Grimsvötn (Iceland), Cordon Caulle (Chile) and Nabro (Ethiopia) have drastically heightened the level of awareness in the general population of how volcanic activity can affect everyday life by disrupting air travel. The ingestion of airborne volcanic matter into jet turbines may cause harm by (1) abrasion of engine parts, (2) destabilisation of the fuel/air mix and its dynamics and (3) by melting and sintering ash onto engine parts. To investigate the behaviour of volcanic ash upon reheating, we have performed experiments at ten temperature steps between 700 and 1600°C on (1) fresh volcanic ash from the final explosive phase of the 2010 Eyjafjallajökull (EYJA) eruption and (2) two standard materials used in ingestion tests in the history of turbine testing (MIL E 5007C test sand, MIL; Arizona Test Dust, ATD). We confirm expected large differences in the samples' response to thermal treatment. We quantify the physical basis for these differences using thermogravimetry and differential scanning calorimetry. Glassy volcanic ash softens at temperatures that are considerably lower than those required for crystalline silicates to start to melt. We find that volcanic ash starts softening at temperatures as low as 600°C and that complete sintering takes place at temperatures as low as 1050°C. Accordingly, the ingestion of volcanic ash in the hot zone of turbines will rather efficiently transform the angular volcanic particles into sticky droplets with a high potential of adhering to surfaces. These experiments demonstrate both a large variability in the material properties of ash from Eyjafjallajökull volcano and a strong contrast to the behaviour of the test sands. In light of these differences, the application in volcanic crises of models of the impact of ash on operability of passenger jet turbines that have been based on test sand calibrations must be re-evaluated. We stress as well that ingestion tests should not only investigate the turbine's response to ash concentration (g/m^3) but also to ash dosage.

BACKGROUND

For many years, the civil aviation authorities have primarily been concerned with localised ash plumes (Guffanti et al. 2010). The year 2010 however marked a dramatic turning point in our perception of associated risks. Beginning with mild explosive activity on the snow-covered eastern flank, the vent position of the Eyjafjallajökull eruption changed to within the glacier-covered summit caldera, causing a marked increase in eruption explosivity (Gudmundsson et al. 2012). The prevailing winds transported the volcanic ash over large areas of the Northern hemisphere and, soon after the eruption onset, caused widespread airport closure in Northern and Central Europe. The dramatic extent of the disruption stemmed from the "zero ash tolerance" guideline followed by decision-makers at the time of the

Eyjafjallajökull eruption. This guideline (Miller and Casadevall 2000; International Civil Aviation Organization [ICAO] 2007) had been implemented by the International Civil Aviation Organization [ICAO] after the 1982 Galunggung (Indonesia) and 1989 Redoubt (USA) incidents (Guffanti et al. 2010; Dunn 2012), based on the Proceedings of the First International Symposium on Volcanic Ash, held in Seattle, Washington, in July 1991 (Casadevall 1994). This rule had been widely accepted by all legal authorities and airline companies, as well as airplane and turbine manufacturers. During the Eyjafjallajökull eruption, the combination of eruption duration, meteorological situation and area of airspace closure lead to substantial economic loss, which was not restricted to airline and airport companies (Budd et al.2011) but also affected industrial activities and goods production.

The increasing duration of airspace closure and the consequent magnitude of economic loss and logistical problems created an increasing reticence towards the imposed flight ban on the part of the airline and cargo companies. However, the rigid implementation of the "zero ash tolerance" guideline undoubtedly avoided hazardous ash encounters. While in-flight ash detection is still a complicated task (Prata and Tupper 2009), airspace contamination by volcanic ash at the time of the Eyjafjallajökull eruption was confirmed for large areas of Europe by direct sampling (Schumann et al. 2011) or LIDAR measurements (Wiegner et al. 2012). At present, the evaluation of the hazard posed by volcanic ash to civil aviation is impeded by 1) the absence of a technology to quantify the ash concentration reliably and quickly at a high temporal and spatial resolution, 2) the invisibility of ash even at concentrations above the currently accepted threshold of 2 mg/m^3(e.g., at night or when overcast; Weinzierl et al. 2012). Recently, the apparent problem of visual detection of volcanic ash (Weinzierl et al. 2012) stimulated a change in terminology and was implemented in the latest ICAO (2013) working paper (IAVWOPSG/7-WP/17).

The "zero ash tolerance" guideline was mainly justified by the limited knowledge about the tolerance of turbine engines to the ingestion of ash particles (Dunn and Wade 1994; Dunn 2012). However, for other types of particulate matter more commonly ingested (e.g., mineral sand) flight operations are permitted within a certain threshold of particle concentration. Under pressure from media and airline companies, turbine manufacturers were asked for a concentration threshold below which safe flying conditions in volcanic ash could be declared possible.

This threshold value, currently replacing the "zero ash tolerance" limit, is set at 2 mg/m^3 (at the time of writing; e.g., Emmott 2010) and has been empirically chosen to fall between known safe flying conditions in mineral sand (data from military jet turbines, e.g., Gabbard et al. 1982) and what are considered 'dangerous' ash concentrations, estimated by reference to ash deposition within turbines during the 1982 and 1989 incidents (the value of ash concentration responsible for the Galungung accident was however largely overestimated; pers. comm. W. Aspinall). A more quantitative assessment of the hazard posed by ash ingestion into turbine engines is highly desirable.

The response of turbine engines to the ingestion of different types of particulate matter will strongly depend on their chemical and physical characteristics, especially at high temperature. Particles suspended in the atmosphere may have very different origins, including volcanic ash, aeolian sand or incineration residues, and thus different chemical and physical characteristics. Particle shape imposes a strong effect on the viscous sintering rates in so far as sintering timescales are dependent on curvature of the surface in contact with the substrate in a jet engine. Therefore, a particle with a high aspect ratio, which impacts the jet engine surface end-on, will sinter at a faster rate than a spherical particle of the same volume because the contact curvature is higher than the spherical equivalent. It arises from sintering theory that particle roughness likely plays a role, albeit a less important one. The chemical composition of rocks is usually declared as a list of elements expressed as oxides. With the exception of natrocarbonatitic melts that also may erupt explosively (Keller et al. 2010), the major constituent in volcanic rocks is silica, SiO_2. The SiO_2 content, however, may reflect the contributions of many silicate phases other than quartz crystals. Furthermore, due to high cooling or degassing rates during eruption, volcanic ash usually contains a fraction of silicate glass, i.e., an amorphous phase lacking long-range crystallographic order. Glass and crystals behave very differently during heating. Glass may soften and melt, deform and stick to surfaces, at temperatures as low as 700°C. Deposition inside turbines will change the internal aerodynamic conditions, affect the temperature of individual components (e.g., by clogging cooling holes), and may react chemically with the thermal-barrier coating. In order to sustain the high temperatures necessary for increased efficiency, parts within the hot zone of the turbine rely on thermal-barrier coatings to operate at temperatures in excess of the melting point of the underlying alloys. Crystalline silicates, in contrast, typically exhibit melting temperatures far higher than the glass softening temperature; crystalline SiO_2, for example, has a melting point above 1700°C. Accordingly, the effect of ash on the operational reliability of aircraft turbines is expected to be very different from that of any mineral sand dominated by quartz.

Sintering of glass, crystals and multi-phase mixtures is a relatively well-understood process in ceramics (Scherer 1977; Scherer and Bachman 1977), physics (Frenkel 1945) and volcanology literature (Sparks et al. 1999). Sintering begins with the formation of necks between particles and involves the transition of a dominantly granular material to a porous framework (Scherer 1977). This initial stage sintering or sticking occurs by viscous neck formation in supercooled melts (high temperature equivalent of glass) and by diffusive neck formation in crystalline material (Frenkel1945). For a given temperature, the latter is generally a slower process but depends strongly on the crystal composition and thermal stability (Zarzycki 1991; Uhlmann et al. 1975). However, in the simpler glass/melt system, is generally accepted that sintering timescales (τ) are dependent on initial grain size (R), melt viscosity (η) and melt surface tension (γ); given by (Uhlmann et al.1975):

$$\tau = \frac{R\eta}{\gamma}$$

(1)

In our experiments with natural volcanic ash, for a constant grain size and composition, the sintering timescale is therefore entirely governed by viscosity and therefore time and temperature dependent. However at the same temperature conditions, the grain size of ingested ash particles should play a critical role in the timescale of effective sintering and therefore grain size will be the prime parameter controlling the probability that particles colliding with each other or with surfaces will adhere and accumulate in engines (Vasseur et al. 2013). Substrate materials in jet engines are not well publicised and therefore the sintering behaviour of particles on a particular substrate remains unconstrained. There is remarkably little literature on sintering as a function of substrate material. This is especially true when considering a single particle and not a granular mixture. However, there is prodigious literature showing and modelling how this can happen in, for example, coal combustion devices with metal substrates

(Tomeczek et al. 2004; Song et al. 2009). For this reason we consider the sintering of particles to other particles of the same material as an approximate proxy to the timescales and characteristics of sintering to a metal substrate or to a thermal barrier coating. We concede that further research in this area is necessary. For the theoretical model (Eq. 1), a spherical particle shape is assumed. Figure 1 clearly shows that none of the three samples is comprised of particles of this shape. The shape of the particles will affect the sintering timescale, as heat conductivity is surface area controlled. The effect of grain shape on sintering timescale was not the target of this study.

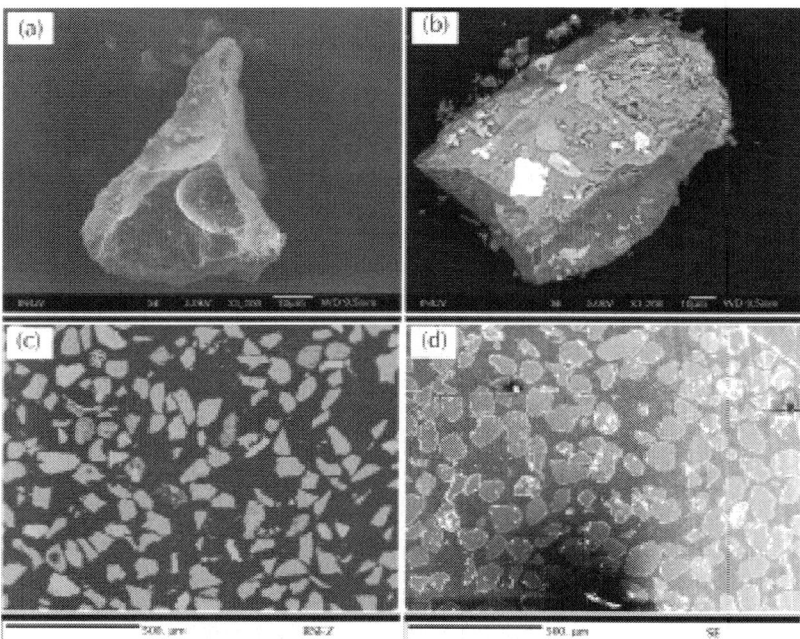

Figure 1: SEM images of single and embedded grains. a) and b)showing different EYJA grains showing the porous nature (a) and the microlite content (b); c) shows the angular shape of ATD grains; d)shows the rounded shape of MIL grains.

Several working groups have characterized Eyjafjallajökull ash (Bonadonna et al. 2011; Dellino et al. 2012; Taddeucci et al. 2011). Further, reaction with metallic surfaces was investigated for natural ash (Gislason et al. 2011) and chemically comparable analogue materials

(Mechnich et al.2011) by constraining "size, shape, and hardness [...] as well as the chemical composition of the salt condensates on the particle surfaces" of ash that was sampled dry. To the best of our knowledge, no investigation has compared the response of Eyjafjallajökull ash and mineral sands to reheating.

In this study, we compare the ash collected during the Eyjafjallajökull eruption (EYJA) to two different test sands commonly employed as turbine contaminants; the "Arizona Test Dust" (ATD) and the "MIL E 5007C" test sands (MIL). We focus on the response of static powder samples to heating for 30, 60 or 120 minutes, respectively, at 1 bar pressure.

Sample Characterisation

Eyjafjallajökull Ash

The ash sample E4 (EYJA) was collected on 18 May 2010 along the plume dispersal axis at 7.5 km north from the vent as it sedimented from the ash cloud. The sample's grain size distribution is smaller than 1 mm and displays a positive skewness (modal value at 500 µm) with a secondary peak at 63 µm. This bimodal distribution can be attributed to the process of aggregation by finer particles (< 125 µm) acting at the time of sampling, enabling the premature depletion of the finer grains from the ash cloud (Taddeucci et al. 2011). Under Field-Emission Scanning Electron Microscope (FE-SEM) the sample appears almost entirely composed of juvenile material, ranging in vesiculation from highly vesicular (Figure 1a) to non-vesicular. Clasts are mostly glassy with variable contents of microlites (crystals smaller than 10 µm, mostly plagioclase, clinopyroxene and oxides, Figure 1b). Chemical bulk composition has been determined by X-Ray Fluorescence (XRF) analysis, while single-spot composition of the glass matrix has been determined by Electron Microprobe Analysis (EMPA, Table 1). Higher values of SiO_2 within the glassy matrix with respect to the bulk composition are consistent with the presence of crystals in the groundmass and, although rare, of loose crystals in the whole sample (Taddeucci et al. 2011).

Table 1: Representative EPMA and XRD results of the EYJA, ATD and MIL samples used in this study

	EPMA point analysis					XRD bulk rock composition			
	A	C	D	E	F	A	C	E	F
	EYJA	MIL	MIL	ATD	ATD	EYJA	MIL	ATD	ATD
	φ < 63 μm	φ < 63 μm	90 < φ < 125 μm	90 < φ < 125 μm	φ < 63 μm	φ < 63 μm	φ < 63 μm	90 < φ < 125 μm	φ < 63 μm
SiO_2	66.38	100.48	100.35	100.13	100.14	60.72	95.92	88.34	74.02
Al_2O_3	15.62	0.03	0.06	0.02	0.02	14.82	2.97	5.67	10.79
FeO	4.81	0.01	0.03	0.02	0.07	8.99	0.46	0.92	2.97
MnO	0.16	0.00	0.00	0.00	0.00	0.21	0.00	0.03	0.09
MgO	0.62	0.00	0.00	0.01	0.00	2.51	0.15	0.47	1.33
CaO	2.78	0.03	0.01	0.02	0.02	4.58	0.04	1.25	2.49
Na_2O	5.26	0.01	0.01	0.00	0.01	5.81	0.22	1.03	1.32
K_2O	2.61	0.01	0.01	0.01	0.00	2.27	1.59	1.78	2.65
TiO_2	0.70	0.01	0.00	0.00	0.00	1.39	0.13	0.12	0.38
P_2O_5	0.18	0.00	0.00	0.00	0.01	0.25	0.03	0.05	0.13
Cl	0.22	0.00	0.00	0.01	0.00	Not analysed			
Total	99.34	100.58	100.47	100.22	100.27	101.55	101.51	99.66	96.17

Sample labeling, measured values and cumulative results in bold for clarification.

Kueppers *et al.*

Kueppers *et al. Journal of Applied Volcanology* **2014** 3:4, doi:10.1186/2191-5040-3-4

Arizona Test Dust

The characteristics of the Arizona Test Dust (ATD) follow the specification of the International Organization for Standardization ISO 12103–1 "Road Vehicles - Test Dust for Filter Evaluation". The ATD is used as a test contaminant for fuel system components, water filter performance evaluation and other custom applications. It consists of a granular material of mixed silicate mineralogy with grain size smaller than 200 μm (Figure 1c). We purchased two different grades of ATD from Powder Technology Inc., ATD-A2 fine grade (< 120 μm) and ATD-A4 coarse grade (< 200 μm), respectively. Grains from both samples are texturally homogeneous, dense and display angular shapes. Chemical compositions of the bulk samples and average composition of single grains have been determined from XRF and EMPA analyses respectively, and are reported in Table 1. Since ATD is mainly employed for the testing of particulate filters, the manufacturer has no criteria for strict controls on its bulk composition and reports the range of compositions for different ATD grades (expressed in wt% of oxides) as follows: SiO_2 68.0–76.0; CaO 2.0-5.0; Al_2O_3 10.0–15.0; MgO 1.0-2.0; Fe_2O_3 2.0–5.0; TiO_2 0.5-1.0; Na_2O 2.0–4.0; K_2O 2.0-5.0. XRF analyses of the bulk sample (Table 1) show that the ATD-A2 composition is in good agreement with the composition given by the manufacturer, while ATD-A4 composition is slightly different, being dominated by SiO_2 with minor wt% of Al_2O_3, FeO, CaO, Na_2O, K_2O.

MIL E 5007C

The characteristics of the sample MIL E 5007C (MIL) follow the specification of the USA Department of Defence standards (also called Military Standard; MIL-STD) for engine, aircraft, turbojet and turbofan. In good commercial quality, it consists of crushed quartz finer than 1 mm. For our experiments, we sieved the bulk sample to obtain two sub-samples, $90 < \phi < 125$ μm (Figure 1d) and $\phi < 63$ μm. BSE images reveal a better rounding of most MIL grains in respect to ATD. Chemical analyses of single grains and bulk sample (Table 1) are in good agreement and confirm the homogeneous mineralogical composition (almost entirely pure quartz) of the samples.

METHODS

Thermal Treatment

Sintering is the process of sticking and cohesion of initially granular particles as a function of heating rate, temperature, composition and grainsize. Sintering is the adherence of particles by diffusive or viscous remobilisation of material to form necks, joining particles in contact with one another. In this study we assess the grainsize- and temperature-dependence of sintering times. We performed 25 experiments with 133 sample specimens at ten temperature steps between 700 and 1600°C and ambient atmosphere. We started with 0.3 g of new material placed in a Calcium Phosphate-crucible (i.e. static experiments). We observed some limited chemical and mechanical interaction between sample and crucible. Chemical interactions are not a focus of the current study; however, the mechanical interaction, the flowing of molten sample into the crucible's pore space, was clearly correlated with the experimental temperature and was constrained qualitatively. Irrespective of their differing chemical compositions, porosity within the crucibles used in this study is also a property exhibited by thermal barrier coatings (see Figures eight and nine in Mechnich et al. 2011). For each sample (EYJA, ATD and MIL) we used two different grain sizes: $\phi < 63$ μm and $90 < \phi < 125$ μm. Sets of 18 samples have been placed in the furnace at the experimental temperature, removed after 30, 60 or 120 minutes, respectively, and allowed to cool under ambient conditions. After cooling, we qualitatively assessed macroscopic changes by classifying the samples in one of the following classes: granular (unsintered powder), partially sintered (individual particles adhere at points of contact and neck formation), efficiently sintered and texturally completely densified. Selected samples were embedded in epoxy, cut and polished to allow thin section and SEM analysis.

Thermal Stability

A series of three different measurements have been performed to quantitatively constrain the sample responses to heating.

- Thermogravimetric (TG) measurements were carried out using a *Netzsch STA 449C* thermal balance. Sample powder of approx. 30 mg was heated in a platinum crucible (with lid) at a heating rate of 10 K/min to 1500°C in an argon atmosphere.

- Calorimetric (differential scanning calorimetry (DSC)) measurements have been performed using a high temperature, low sensitivity *Netzsch STA 449C*. Approx. 30 mg were heated in a platinum crucible with a heating rate of 10 K/min to 1325°C.

- Calorimetric (differential scanning calorimetry, DSC) measurements have been performed using a low temperature, high sensitivity *Setaram Sensys Evo*. Approximately 60 mg were heated in a platinum crucible with a heating rate of 10 K/min to 800°C in an argon atmosphere. This measurement was performed only on EYJA samples.

RESULTS

- The major element composition (Table 1) was quantified as single point (microprobe) and bulk rock (XRF) analyses. FE-SEM images reveal a variable microlite content for EYJA (Figure 1a,b) while ATD (Figure 1c) and MIL (Figure 1d) are almost exclusively composed of quartz grains. In the ATD sample, we observe that the quartz grains are coated with another mineral phase, most likely clay (see DSC results). The differences between the two chemical data sets are minor for MIL, moderate for EYJA and high for ATD. We observe a generally angular shape for clasts of EYJA and ATD while most clasts of MIL are subrounded to rounded. EYJA clasts show abundant segments of bubble walls from broken bubbles. The EYJA ash was found to be strongly degassed with measured water contents below 0.1 wt.%.

- Sintering experiments: samples were held in the furnace for 30, 60 or 120 minutes, respectively, and allowed to cool under ambient conditions (Figure 2a). Most samples changed colour due to oxidizing conditions within the furnace. The samples were investigated for particle-particle cohesion by using a needle to test the possibility of low-force indentation. We defined three categories; loose, sintered and homogenised. We find that the starting grain size has a minor influence during experiments on EYJA samples and that macroscopically detectable sintering starts consistently between 850 and 900°C (Figure 2b). At higher temperatures (Figure 2c), finer samples show a more complete densification. During experiments with crystalline sands, the influence of the grain size is more noticeable and sintering was found to occur at 1100°C (Figure 2c) for fine ATD and 1200°C for coarse ATD. We observed complete homogenisation at 1050°C for the volcanic ash; at 1200°C (Figure 2d) for the fine ATD; and at 1400°C for the coarse ATD. The MIL sample shows signs of a homogenisation onset at 1400°C (Table 2).

- Infiltration of substrate: Above 850°C, all volcanic ash samples show signs of sintering, the extent of which correlated positively with experimental temperature. Samples from three experiments (60 min at T_{exp} = 950, 1000 and 1050°C) were impregnated and sectioned vertically to expose the reaction rim between sample and sample holder. This analysis mainly aims at understanding the physics of the process and not the inevitable chemical reactions occurring between the parts of the sample in contact with the crucible. We observed the strong influence of viscosity on the infiltration of silicate melt into the porous structure of the used crucibles. At 950°C, no significant melting is obvious (Figure 3a), best evidenced at higher resolution through the newly-formed glass necks (Figure 3b). At 1000°C, the glass necks become more numerous and thicker (Figure 3c) but only the basal ash grains show efficient loss of inter-particle porosity by viscous flow (Figure 3d). At 1050°C, all ash particles have formed a coherent and dense mass (Figure 3e) that was low enough in viscosity to infiltrate the sample holder (Figure 3f).

Figure 2: Photos of the samples (a) after 60 min at 700°C (no observable change), (b) after 60 min at 900°C, (c) after 60 min at 1100°C and (d) after 60 min at 1200°C (each crucible has a diameter of 22 mm). Arrangement of samples in (a) through (c): top row from left to right: EYJA, ATD and MIL (all $\phi < 63$ µm); bottom row from left to right: EYJA, ATD and MIL (all $90 < \phi < 125$ µm). Arrangement of samples in (d): top row from left to right: ATD and MIL (all $\phi < 63$ µm); bottom row from left to right: ATD and MIL (all $90 < \phi < 125$ µm). After 60 minutes at 700°C (a), all samples are still loose powders. After 60 minutes at 900°C(b), EYJA samples show signs of sintering (crusted surface), all other samples are still loose powders. After 60 minutes at 1100°C (c), EYJA samples are completely sintered whereas all other samples are still loose. After 60 minutes at 1200°C (d), ATD is partially sintered ($90 < \phi < 125$ µm) and completely welded ($\phi < 63$ µm), respectively, whereas MIL samples are still loose powders.

Table 2: Qualitative results of the degree of sintering of the EYJA, MIL and ATD samples

| | Experimental temperature (°C) | | | | | | | | | | |
		700	800	850	900	950	1000	1050	1100	1200	1400	1600
Sample	EYJA (Φ <63 μm)	No	No	Onset	Yes	Yes	Yes	Texturally densified	n.p.	n.p.	n.p.	n.p.
	EYJA (90 μm < Φ < 125 μm)	No	No	Onset	Yes	Yes	Yes	Texturally densified	n.p.	n.p.	n.p.	n.p.
	MIL (Φ <63 μm)	No	No	No	No	No	No	No	No	No	Onset	Yes
	MIL (90 μm < Φ < 125 μm)	No	No	No	No	No	No	No	No	No	Onset	Yes
	ATD (Φ <63 μm)	No	No	No	No	No	No	No	Onset	Yes	Texturally densified	n.p.
	ATD (90 μm < Φ < 125 μm)	No	No	No	No	No	No	No	No	Onset	Yes	n.p.

Abbreviations: n.p. experiment not performed.

Kueppers *et al.*

Kueppers *et al. Journal of Applied Volcanology* **2014** 3:4, doi:10.1186/2191-5040-3-4

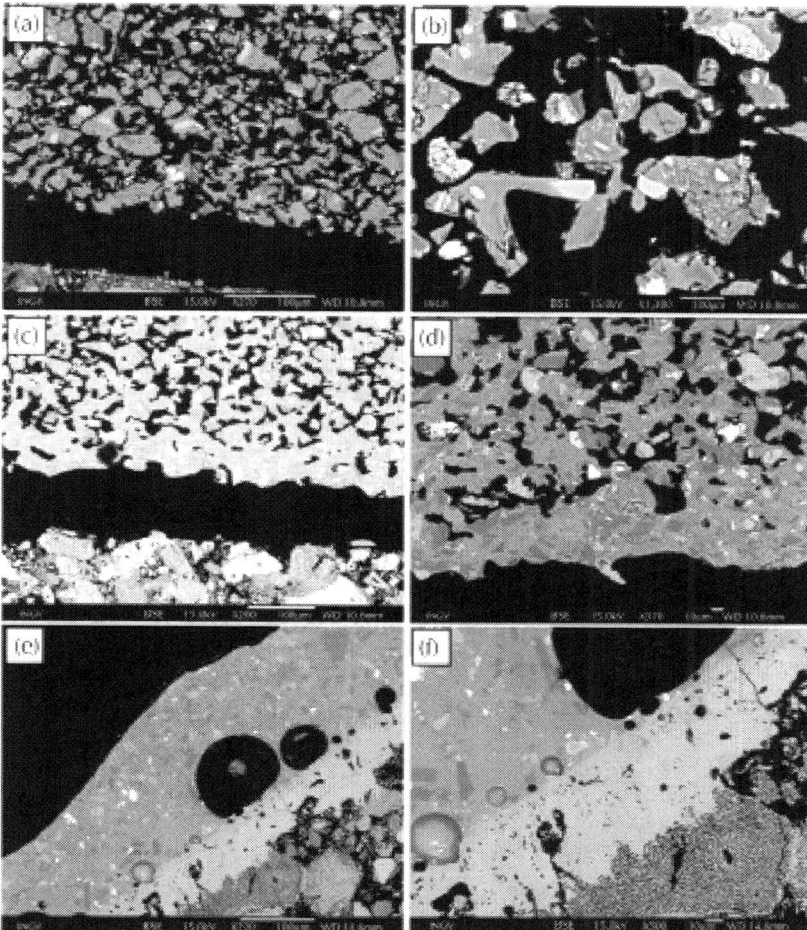

Figure 3: SEM images of impregnated thin section showing the different degrees of sintering from quasi-initial conditions after 60 minutes at 950°C (a overview, b detail), through considerable neck formation after 60 minutes at 1000°C (c overview, d detail) to complete textural densification after 60 minutes at 1050°C (e overview, f detail).

- TG: The thermogravimetric measurements revealed (independent of grainsize) no mass loss for EYJA and MIL in the temperature interval between room temperature and 1325°C (experimental error +/− 0.1 wt.%). In contrast, ATD reveals a significant mass loss of up to 3 wt.% between room temperature and 700°C (Figure 4).

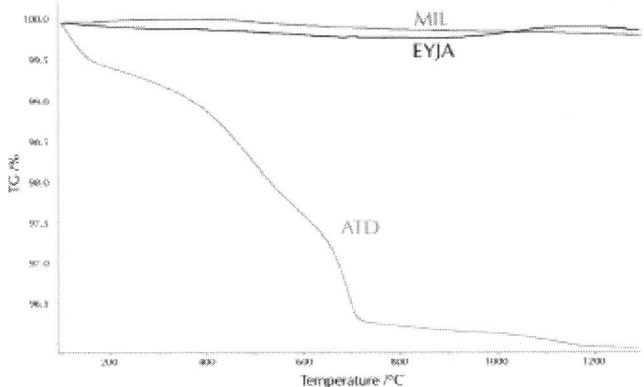

Figure 4: Results of thermogravimetry analysis (Pt crucibles were used at 10 K/min heating rate under argon atmosphere.) for the $90 < \phi < 125$ μm grainsize fraction. EYJA and MIL samples show a fairly similar behaviour whereas ATD shows a significant weight loss (4 wt.%) that can be attributed to decomposition reactions, most likely of H_2O-bearing phases.

- The heat flow signal measured by high-T DSC across a heating interval up to 1325°C shows a fundamental difference between EYJA (black line) and the other samples (Figure 5); the latter two showing a unique peak at 573 +/− 2°C (α to β-quartz transition).

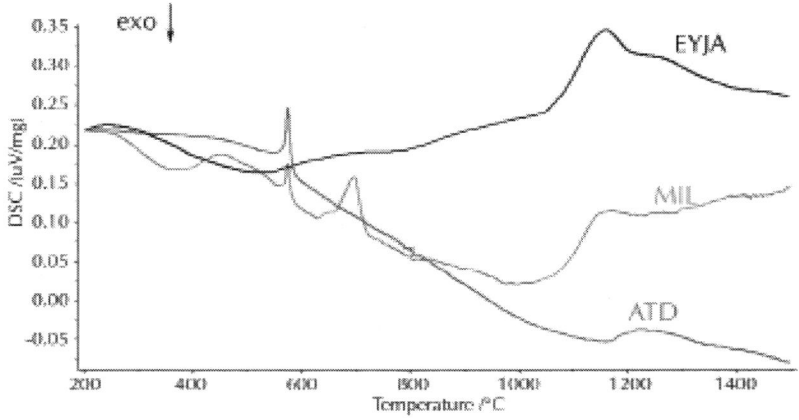

Figure 5: Low-res DSC results (Pt crucibles were used at 10 K/min heating rate under argon atmosphere.) for the $90 < \phi < 125$ μm grainsize fraction.

Only ATD and MIL show the α-β-quartz transition at 573°C. MIL shows further secondary peaks between 400 and 700°C manifesting the decomposition of secondary phases. The EYJA sample shows a slight bump between 600 and 750°C. This bump corresponds to the relaxation of the glass phase upon heating (= glass transition interval) and is shown in a more pronounced way in Figure 6.

- The glass transition interval was measured for EYJA by low-T DSC and revealed a broad, endothermic peak starting at 600°C (Figure 6). Remelted ash from the TG measurements shows are more pronounced peak, also starting at 600°C. Here, the glass structure starts to relax, manifested in a measurable change in sample length (Figure 4). We did not observe a comparable response of MIL or ATD.

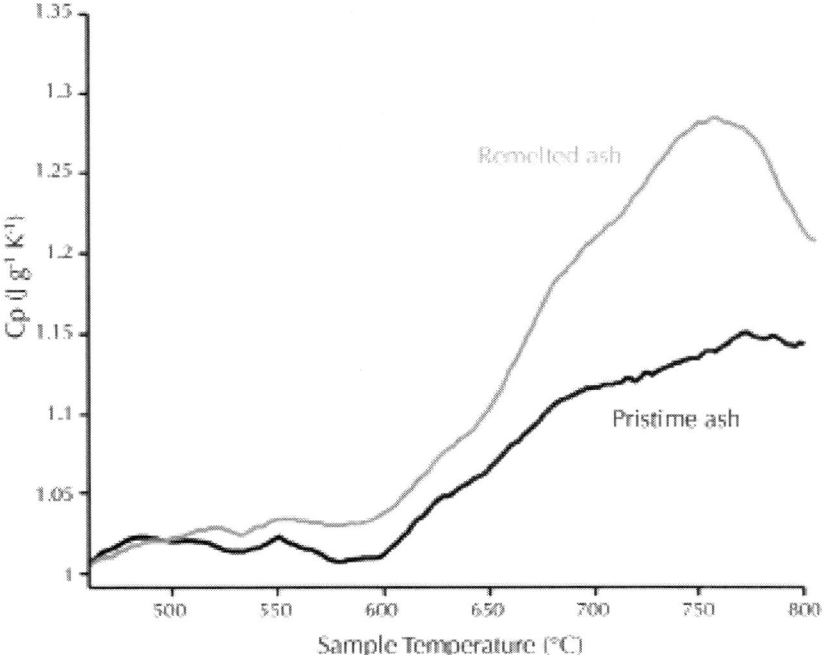

Figure 6: High-resolution DSC results (Platinum crucibles were used at 10 K/min heating rate under argon atmosphere.) for the $90 < \phi < 125$ µm grainsize fraction of EYJA. The black curve was achieved measuring a pristine sample as collected in the field. The grey curve was achieved measuring a remelted sample after the thermogravimetry experiment (Figure 5). The peak at 750°C

can be attributed to the glass transition. The peak of the pristine ash is less pronounced due to the crystals content in the ash grains. The curve of "remelted ash" shows the minor difference between bulk rock and interstitial melt.

DISCUSSION AND CONCLUSIONS

Constraining the impact of volcanic ash on the operational reliability of passenger turbines is difficult because many important parameters are highly variable due to natural and/or engineering reasons: 1) Volcanic ash composition, 2) glass fraction (Kennedy and Russell 2012), 3) interstitial melt/bulk rock composition, 4) ash concentration, 5) grainsize distribution at turbine inlet, 6) air flow speed inside a turbine, 7) temperature inside a turbine, 8) maximum pressure inside a turbine, 9) degree of grainsize reduction inside the turbine and 10) dominant grainsize in the hot zone of the turbine.

In May 2010, a bubbly magma (Figure 1a) with some crystalline content was erupted at Eyjafjallajökull. The widespread ash affected aviation significantly as modelled results predicted ash over large areas of northern and central Europe. There was no general consensus about 1) the ash concentration, 2) the effective threat posed by volcanic ash, 3) the necessity of air space closure and 4) how to be better prepared in the case of a similar future scenario.

Our experiments demonstrate a significant difference in material properties between volcanic ash from Eyjafjallajökull volcano (EYJA) and the two mineral sands (ATD and MIL). In turbines, the ingested matter is subject to rapid heating. Volcanic ash usually contains a significant fraction of glass whose composition commonly ranges from 40 to 80 wt.% SiO_2. This value should not be misinterpreted as representing quartz. Table 1 shows that MIL is nearly pure quartz, while ATD is mainly comprised of quartz grains coated with secondary phases. These phases break down during heating and release volatiles, leading to a reduction of sample weight (Figure 4). EYJA does not contain any quartz as only ATD and MIL results show a peak at 573 +/− 2°C, indicative of the α-β-transition of the quartz structure (Figure 5).

When heated, glass requires significantly less energy than crystalline matter to change from solid to liquid. Glass does not melt, it softens and transitions to a liquid state. This "glass transition" range was constrained to be between 600 and 800°C at a heating rate of 10 K/min for EYJA (Figure 6). This result differs strongly from results achieved on experimentally generated samples of the same composition (Mechnich et al. 2011). We postulate that this difference lies in the cooling history of our natural samples and not in the composition alone. The glass transition peak is less pronounced than for pure glasses; we relate this to the presence of crystals in the glass and a distribution of iron in different oxidation states (Kremers et al. 2012).

As the viscosity of the EJYA samples below 800°C is very high, no significant flow was possible on experimental timescales (up to 2 hours) and the onset of macroscopic sintering had only been recognized for experiments at >850°C. Figure 3 shows that temperatures of 1000°C needed to be achieved in order to allow for significant neck formation (Figure 3b) and 1050°C to achieve values of viscosity low enough to allow for considerable viscous flow during the timescale of the experiments (Figure 3c). When ingested in the hot zone of turbines, where significantly higher temperatures can be reached, volcanic ash particles are quickly transformed into liquid droplets that have a high probability of adhering to surfaces. In order to adhere (and not bounce off) at high impact velocities (higher than the travel speed of the airplane), the viscosity of the melt must be low enough. Viscosities are strongly dependent on temperature (turbine operation dependent), SiO_2 content (volcano dependent) and the amount of volatiles still dissolved in the melt (pressure dependent). Pressures in the hot zone of turbines can be as high as 300 bar (A. Durant, pers. comm.) and could prevent an outgassing of the melts, thereby keeping the viscosities low (Hess and Dingwell 1996).

Based on our empirical findings from static experiments, we calculated the viscosity of a silicate melt (EYJA interstitial melt composition) as a function of temperature and chemical composition according to Giordano et al. (2008) and constrained 10^8 Pa s as the critical value of viscosity to achieve sintering in our experiments. This corresponds to 880°C. For ATD and MIL, 1100 and 1300°C, respectively, have to be achieved for similar values of viscosity (based on bulk rock composition). We stress that these are most likely minimum temperature values, as higher deformation rates upon impact of moving particles onto surfaces in the turbines will require lower values of viscosity to permit sticking (Wieland et al. 2012). The melting point of pure quartz is around 1700°C and is usually not achieved in passenger jet turbines. As a consequence, ingested MIL will not affect the turbines thermally. In multiphase systems, the bulk melting point is lowered to the eutectic temperature. The investigated ATD is comprised by quartz with up to 23 wt.% of other phases. These phases break down during heating (but before 700°C) and release volatiles. Possible phases are clay minerals (releasing H_2O) and/or calcite (releasing CO_2). As a consequence, melting was observed for ATD at 1200°C.

Adhering melt droplets will affect the flow of cooling air through the turbine, both as a whole and over individual components. Additionally, the silicate melt may interact chemically (not part of this study) and mechanically with the thermal barrier coating (TBC) on component surfaces in the hot zone of the turbine. At low viscosity, the melt will flow into the intercrystalline pore space of the TBC. Upon cooling, differential contraction due to differences in thermal expansivity of glass and the TBC, may cause damage to the thermal barrier coatings.

We still face difficulties in quantifying the concentrations of volcanic ash in a small volume of air at high temporal and spatial reliability (Sears et al. 2013). Thus, ash ingestion tests should be performed to constrain the effect of ash concentration and total load. We stress that *fresh volcanic ash* should be used in order to avoid a bias of the results due to alteration/weathering effects. Certainly, ingestion tests with standard sands cannot be reliable proxies for the thermal risk of volcanic ash in turbines. We are convinced that concentration thresholds alone fail to account for the possible impact of ingested volcanic ash on the operational reliability of passenger jet turbines. We propose the introduction of an ash dose threshold as the impact of volcanic ash on the operational reliability of passenger jet turbines is the result of 1) ash concentration (mg/m^3), 2) the amount of air ingested per second (up to hundreds of m^3, dependent on flight situation) and 3) the time span an airplane will possibly travel in ash-contaminated air space.

AUTHORS' CONTRIBUTIONS

UK and CC performed and analysed the sintering experiments, KUH the thermogravimetry and DSC analysis. UK, CC and JT investigated the samples before and after the experiments with EMPA and SEM.

KUH and FBW reviewed the sintering literature. UK, CC, KUH and DBD discussed the implications of the results. UK, CC, FBW and JT drafted the manuscript. All authors have read and approved the final manuscript.

ACKNOWLEDGEMENTS

We thank Willy Aspinall, Paul Ayris, Adam Durant, Marianne Guffanti and Larry Mastin for vivid discussions and fruitful insights. We thank Soraya Heuss-Aßbichler for the XRF measurements and Hilger Lohringer for sample preparation. DBD wishes to acknowledge the support of a Research Professorship LMUexcellent of the Bundesexzellenzinitiative as well as Advanced Grant "EVOKES" 247076 of the ERC. The AXA Research Fund Project "Risk from volcanic ash in the Earth system" has primarily supported this work. We thank the two anonymous reviewers for their input.

REFERENCES

1.   Bonadonna C, Genco R, Gouhier M, Pistolesi M, Cioni R, Alfano F, Hoskuldsson A, Ripepe M (2011) Tephra sedimentation during the 2010 Eyjafjallajökull eruption (Iceland) from deposit, radar, and satellite observations. J Geophys Res 116:B12202 doi:10.1029/2011JB008462

2.   Budd L, Griggs S, Howarth D, Ison S (2011) A fiasco of volcanic proportions? Eyjafjallajökull and the closure of european airspace. Mobilities 6(1):31-40

3.   Casadevall TJ (1994) Volcanic ash and Aviation Safety; Proceedings of the First International Symposium on Volcanic Ash and Aviation Safety Held in Seattle. Washington: U.S. Geological Survey Bulletin. p 2047 in July 1991

4.   Dellino P, Gudmundsson MT, Larsen G, Mele D, Stevenson JA, Thordarson T, Zimanowski B (2012) Ash from the Eyjafjallajökull eruption (Iceland): Fragmentation processes and aerodynamic behaviour. J Geophys Res 117:B00C04 doi:10.1029/2011JB008726

5. Dunn MG (2012) Operation of gas turbine engines in an environment contaminated with volcanic ash. J Turbomach 134(5): :051001-051001-18, doi: 10.1115/1.4006236

6. Dunn MG, Wade DP (1994) Influence of volcanic ash clouds on gas turbine engines. In casadevall TJ volcanic ash and aviation safety; proceedings of the first international symposium on volcanic Ash and aviation safety held in Seattle, Washington, in July 1991: U.S. Geol Surv Bull 2047:107-117

7. Emmott P (2010) Eyjafjallajökull – the impact of volcanic ash on aircraft engines. Oral commun Atlantic Conf Eyjafjallajökull Aviation Keflavik. http://en.keilir.net/static/files/Aviation/PDF/Eyjafjallajokull_and_Aviation_Conference_Program.pdf

8. Frenkel J (1945) Viscous flow of crystalline bodies under the action of surface tension. J Phys 9(5):385-391

9. Gabbard CB, LeLevier RE, Parry JFW (1982) Dust-Cloud Effects on Aircraft Engines – Emerging Issues and new Damage Mechanisms. A Case Study of a Mt. St. Helens Experience and its Implications for Nuclear-Weapon-Lofted Dust-Cloud Effects. US Defense Nuclear Agency Report DNA-TR-82-18. http://www.dtra.mil/documents/foia/DNA-TR-82-18.pdf

10. Giordano D, Russell JK, Dingwell DB (2008) Viscosity of magmatic liquids: a model. Earth Plan Sci Lett 271:123-134

11. Gislason SR, Hassenkam T, Nedel S, Bovet N, Eiriksdottir ES, Alfredsson HA, Hem CP, Balogh ZI, Dideriksen K, Oskarsson N, Sigfusson B, Larsen G, Stipp SLS (2011) Characterization of eyjafjallajökull volcanic ash particles and a protocol for rapid risk assessment. Proc Nat Acad Sci 108(18):7307-7312

12. Gudmundsson MT, Thordarson T, Höskuldsson A, Larsen G, Björnsson H, Prata FJ, Oddsson B, Magnússon E, Högnadóttir T, Petersen GN, Hayward CL, Stevenson JA, Jónsdóttir I (2012) Ash generation and distribution from the april-may 2010 eruption of eyjafjallajökull. Iceland Sci Rep 2:572 10.1038/srep00572

13. Guffanti M, Casadevall TJ, Budding K (2010) Encounters of Aircraft with Volcanic ash Clouds; A Compilation of Known Incidents, 1953–2009. U.S. Geological Survey Data Series 545. http://pubs.usgs.gov/ds/545/DS545.pdf

14. Hess KU, Dingwell DB (1996) Viscosities of hydrous leucogranitic melts: a non-arrhenian model. Am Min 81:1297-1300

15. International Civil Aviation Organization [ICAO] (2007) Manual on Volcanic ash, Radioactive Material and Toxic Chemical Clouds. International Civil Aviation Organization Doc 9691– AN/954. http://www.paris.icao.int/news/pdf/9691.pdf

16. International Civil Aviation Organization [ICAO] (2013) IAVWOPSG/7-WP/17, International Airways Volcano Watch Operations GROUP, 7th Meeting, Follow-up of IVATF Recommendation 4/10m – Definitions of visible ash and discernable ash for operational use. accessed May 22, 2013,http:// www.icao.int/safety/meteorology/iavwopsg/IAVWOPSG%20 Meetings%20Meadata/IAVWOPSG.7.WP.017.5.pdf

17. Keller J, Klaudius J, Kervyn M, Ernst GGJ, Mattson HB (2010) Fundamental changes in the activity of the natrocarbonatite volcano Oldoinyo Lengai, Tanzania. Bull Volcanol 72:893-912

18. Kennedy LA, Russell JK (2012) Cataclastic production of volcanic ash at Mount Saint Helens. Phys Chem Earth 45–46:40-49

19. Kremers S, Lavallee Y, Hanson J, Hess KU, Chevrel MO, Wassermann J, Dingwell DB (2012) Shallow magma-mingling-driven Strombolian eruptions at Mt. Yasur Volcano, Vanuatu. Geophys Res Lett 39: L21304

20. Mechnich P, Braue W, Schulz U (2011) High-temperature corrosion of EB-PVD yttria partially stabilized zirconia thermal barrier coatings with an artificial volcanic Ash overlay. J Am Cer Soc 94:925-931

21. Miller TP, Casadevall TJ (2000) Volcanic ash Hazards to Aviation. In: Sigurdsson H (ed) Encyclopedia of Volcanoes, San Diego: Academic. pp 915-930

22. Prata AJ, Tupper A (2009) Aviation hazards from volcanoes: the state of the science. Nat Haz 186:91-107

23. Scherer GW (1977) Sintering of low-density glasses: I, theory. J Am Cer Soc 60(5–6):236-239

24. Scherer GW, Bachman DL (1977) Sintering of low-density glasses: II, experimental study. J Am Cer Soc 60(5–6):239-243

25. Schumann U, Weinzierl B, Reitebuch O, Schlager H, Minikin A, Forster C, Baumann R, Sailer T, Graf K, Mannstein H, Voigt C, Rahm S, Simmet R, Scheibe M, Lichtenstern M, Stock P, Ru ba H, Schäuble D, Tafferner A, Rautenhaus M, Gerz T, Ziereis H,

Krautstrunk M, Mallaun C, Gayet JF, Lieke K, Kandler K, Ebert M, Weinbruch S, Stohl A, Gasteiger J, Gross S, Freudenthaler V, Wiegner M, Ansmann A, Tesche M, Olafsson H, Sturm K (2011) Airborne observations of the Eyjafjalla volcano ash cloud over Europe during air space closure in April and May 2010. Atmos. Chem. Phys 11:2245-2279 http://dx.doi.org/10.5194/acp-11-2245-2011

26. Sears TM, Thomas GE, Carboni E, Smith AJA, Grainger RG (2013) SO$_2$ as a possible proxy for volcanic ash in aviation hazard avoidance. J Geophys Res (Atmospheres) 118(11): doi:10.1002/jgrd.50505

27. Song W, Tang L, Zhu X, Wu Y, Rong Y, Zhu Z, Koyama S (2009) Fusibility and flow properties of coal ash and slag. Fuel 88(2):297-304

28. Sparks RSJ, Tait SR, Yanev Y (1999) Dense welding caused by volatile resorption. J Geol Soc 156(2):217-225

29. Taddeucci J, Scarlato P, Montanaro C, Cimarelli C, Del Bello E, Freda C, Andronico D, Gudmudsson MT, Dingwell DB (2011) Aggregation-dominated ash settling from the eyjafjallajökull volcanic cloud illuminated by field and laboratory high-speed imaging. Geol 39:891-894

30. Tomeczek J, Palugniok H, Ochman J (2004) Modelling of deposits formation on heating tubes in pulverized coal boilers. Fuel 83(2):213-221

31. Uhlmann D, Klein L, Hopper R (1975) Sintering, crystallization, and breccia formation. Moon 13(1–3):277-284

32. Vasseur J, Wadsworth FB, Lavallee Y, Hess KU, Dingwell DB (2013) Viscous sintering: timescales of viscous densification and strength recovery. Geophys Res Lett 40:5658-5664

33. Weinzierl B, Sauer D, Minikin A, Reitebuch O, Dahlkötter F, Mayer B, Emde C, Tegen I, Gasteiger J, Petzold A, Veira A, Kueppers U, Schumann U (2012) On the visibility of airborne volcanic ash and mineral dust from the pilot's perspective in flight. Phys Chem Earth A 45–46:87-102

34. Wiegner M, Gasteiger J, Gross S, Schnell F, Freudenthaler V, Forkel R (2012) Characterization of the Eyjafjallajökull ash-plume: Potential of lidar remote sensing. Phys Chem Earth 45-46:79-86

35. Wieland C, Kreutzkam B, Balan G, Spliethoff H (2012) Evaluation, comparison and validation of deposition criteria for numerical simulation of slagging. Appl Ener 93:184-192

36. Zarzycki J (1991) Glasses and amorphous solids (Vol 9). In: Cahn RW, Hassen P, Kramer EJ (eds) Materials Science and Technology: a comprehensive treatment, VCH Verlagsgesellschaft, Weinheim: Wiley.

Chapter 5

Optimum Parametric Performance Characterization of an Irreversible Gas Turbine Brayton Cycle

Maher M Abou Al-Sood, Kassem K Matrawy, and
Yousef M Abdel-Rahim

Department of Mechanical Engineering, Assiut University, Assiut 71516, Egypt

ABSTRACT

A general mathematical model is developed to specify the performance of an irreversible gas turbine Brayton cycle incorporating two-stage compressor, two-stage gas turbine, intercooler, reheater, and regenerator with irreversibilities due to finite heat transfer rates and pressure drops. Ranges of operating parameters resulting in optimum performance (i.e., $\eta_I \geq 38 \geq \eta_{II} \geq 60\%$, ECOP ≥ 1.65, $x_{loss} \leq 0.150$ MJ/kg, BWR ≤ 0.525, $w_{net} \geq 0.300$ MJ/kg, and $q_{add} \leq 0.470$ MJ/kg) are determined and discussed using the Monte Carlo method. These operating ranges are minimum cycle temperature ranges between 302 and 315 K, maximum cycle temperature ranges between 1,320 and 1,360 K, maximum cycle pressure ranges between 1.449 and 2.830 MPa, and conductance of the heat exchanger ranges between 20.7 and 29.6 kW/K. Exclusive effect of each of the operating parameters on each of the performance parameters is mathematically given in a general formulation that is applicable regardless of the values of the rest of the operating parameters and under any condition of operation of the cycle.

BACKGROUND

First gas turbines developed in the 1930's used to have representative simple cycle efficiencies of about 17% due to low compressor and turbine efficiencies and low turbine inlet temperatures for material stress and thermal limitations. Efforts to improve these efficiencies have specifically or concurrently concentrated in three areas: (1) modifying the working cycle, (2) increasing turbine inlet temperature, and (3) enhancing the performance of cycle components. Recently, developments in material science allow using turbine inlet temperatures up to 1,500°C (i.e., general electric uses a turbine inlet temperature of 1,425°C). Also, continuous modifications of Brayton cycle to include regeneration [1,2], isothermal heat addition [3-6], intercooled compression [7,8], reheat expansion [9,10], and combined modifications [11-14] have resulted in practically doubling the cycle efficiencies. This is because intercooling and reheating result in decreasing the average temperature at which heat is added. Finally, computer-aided design and simulation studies have enabled optimization of cycle components such as compressors and turbines.

The Brayton cycle, as a model of gas turbine power plants, has been optimized for entropy generation [15,16], reversible work [17,18], power [19-22], power density [23-25], internal irreversibilities of compressors and turbines [26,27], pressure drops in heaters, coolers, and regenerators [19,23,24,28], and external irreversibilities of coupling to external heat reservoirs or heat exchangers [20].

Most of the abovementioned literature studies have been carried out to improve the performance of real gas power plants through the optimization of design and operating parameters such as compressor and turbine inlet temperatures, pressure ratios of intercooling, reheat, and conductance of heat exchangers [12,29-33]. However, most of the previously published results found in the open literature are typically specific and valid only for the condition and parameter values taken into consideration in these studies. This means that according to the authors' knowledge, there is no general optimized work that has been done before. Therefore, and for the sake of generalized tackling of this issue, the main objective of the present study of an irreversible regenerative intercooled reheat gas turbine Brayton cycle is to identify the ranges of all design and operating parameters for optimized performance. The design and operating parameters include inlet temperatures to compressors and turbines and pressure ratios of intercooler and reheater. The performance parameters include the first and second law efficiencies, ecological coefficient of performance, back work ratio, exergy losses, network, and heat added.

METHODS

Mathematical Model

Consider a constant mass flow rate, $\dot{m}$, of air, as an ideal gas passing through the gas turbine cycle illustrated in Figures 1 and 2. The cycle can be characterized as follows:

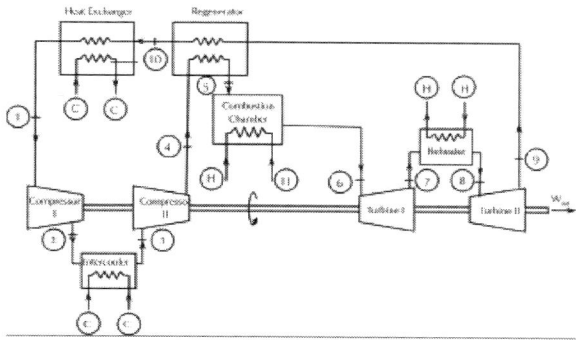

Figure 1: Schematic diagram of a realistic irreversible, regenerative, and reheat Brayton cycle.

- Air is compressed from state 1 to state 4 by two non-isentropic low pressure (LP) and high pressure (HP) compressors with efficiencies, η_{c12} and η_{c34}, and a non-isobaric counter-flow intercooler with effectiveness, ε_{int}. The inlet temperature to the HP compressor is 5% higher than that of the LP compressor. The describing equations for these processes (e.g., [34,35]) are as follows:

$$\eta_{c12} = \frac{w_{c12s}}{w_{c12}} = \frac{h_{2s} - h_1}{h_2 - h_1} \tag{1}$$

$$\eta_{c34} = \frac{w_{c34s}}{w_{c34}} = \frac{h_{4s} - h_3}{h_4 - h_3} \tag{2}$$

$$\varepsilon_{int} = \frac{\dot{Q}_{23}}{\dot{Q}_{int\,max}} = \frac{(UA)_{int}(\Delta T_{LM})_{int}}{\dot{Q}_{int\,max}}$$
$$= \frac{\dot{Q}_{23}}{\min(\dot{C}_W, \dot{C}_{23}) \times (T_2 - T_{C2})} \tag{3}$$

Quantities $\dot{Q}_{23}$, $\dot{C}_W$, and $\dot{C}_{23}$ represent the rate of heat release and heat capacity rates for cooling fluid and air, respectively. The intercooler logarithmic mean temperature difference $(\Delta T_{LM})_{int}$ is defined as follows:

$$(\Delta T_{LM})_{int} = \frac{(T_2 - T_{C3}) - (T_3 - T_{C2})}{\ln((T_2 - T_{C3})/(T_2 - T_{C3}))} \tag{4}$$

- Air is preheated from state 4 to state 5 in a regenerative counter-flow heat exchanger (that will be discussed later in the heat rejection process) and then heated up to a maximum temperature, T_6, by a counter-flow heat exchanger having a rate of heat addition $\dot{Q}_{56}$, an effectiveness ε_{add}, and a logarithmic mean temperature difference $(\Delta T_{LM})_{add}$ defined as follows:

$$\varepsilon_{add} = \frac{\dot{Q}_{56}}{\dot{Q}_{add\,max}} = \frac{(UA)_{add}(\Delta T_{LM})_{add}}{\dot{Q}_{add\,max}}$$
$$= \frac{\dot{Q}_{56}}{\min(\dot{C}_W, \dot{C}_{56}) \times (T_5 - T_{H5})} \tag{5}$$

$$(\Delta T_{LM})_{add} = \frac{(T_5 - T_{H6}) - (T_6 - T_{H5})}{\ln((T_5 - T_{H6})/(T_6 - T_{H5}))} \tag{6}$$

- Air is expanded from state 6 to final state 9 by two non-isentropic LP and HP turbines with efficiencies ηt_{67} and ηt_{89} and one non-isobaric reheater having a rate of heat added, an effectiveness, and a logarithmic mean temperature difference as Q_{78}, ε_{reh}, and $(\Delta T_{LM})_{reh}$. The inlet temperature to LP turbine is 5% lower than that of the HP turbine. The governing equations for these processes are as follows:

$$\eta_{t67} = \frac{w_{t67}}{w_{t67s}} = \frac{h_6 - h_7}{h_6 - h_{7s}} \tag{7}$$

$$\eta_{t89} = \frac{w_{t89}}{w_{t89s}} = \frac{h_8 - h_9}{h_8 - h_{9s}} \tag{8}$$

$$\varepsilon_{reh} = \frac{\dot{Q}_{78}}{\dot{Q}_{reh\,max}} = \frac{(UA)_{reh}(\Delta T_{LM})_{reh}}{\dot{Q}_{reh\,max}}$$

$$= \frac{\dot{Q}_{78}}{\min(\dot{C}_H . \dot{C}_{78}) \times (T_{H7} - T_7)} \tag{9}$$

$$(\Delta T_{LM})_{reh} = \frac{(T_{H7} - T_8) - (T_{H8} - T_7)}{\ln((T_{H7} - T_8)/(T_{H8} - T_7))} \tag{10}$$

- In the heat rejection process 9 to 1 between the exit of HP turbine and inlet of LP compressor, air is firstly cooled in the regenerator (with rate of heat added, effectiveness, and logarithmic mean temperature difference of Q_{45}, ε_{reg}, and $(\Delta T_{LM})_{reg}$, respectively) and finally cooled to state 1 in a counter-flow heat exchanger of parameters Q_{101}, ε_{rej}, and $(\Delta T_{LM})_{rej}$. The governing equations are as follows:

$$\varepsilon_{\text{reg}} = \frac{\dot{Q}_{45}}{\dot{Q}_{\text{reg max}}} = \frac{(UA)_{\text{reg}}(\Delta T_{\text{LM}})_{\text{reg}}}{\dot{Q}_{\text{reg max}}}$$

$$= \frac{\dot{Q}_{45}}{\min(\dot{C}_{45}, \dot{C}_{910}) \times (T_9 - T_4)}$$

(11)

$$(\Delta T_{\text{LM}})_{\text{reg}} = \frac{(T_9 - T_5) - (T_{10} - T_4)}{\ln((T_9 - T_5)/(T_{10} - T_4))}$$

(12)

$$\varepsilon_{\text{rej}} = \frac{\dot{Q}_{101}}{\dot{Q}_{\text{rej max}}} = \frac{(UA)_{\text{rej}}(\Delta T_{\text{LM}})_{\text{rej}}}{\dot{Q}_{\text{bur max}}}$$

$$= \frac{\dot{Q}_{101}}{\min(\dot{C}_W, \dot{C}_{101}) \times (T_{10} - T_{C10})}$$

(13)

$$(\Delta T_{\text{LM}})_{\text{rej}} = \frac{(T_{10} - T_{C1}) - (T_1 - T_{C10})}{\ln((T_{10} - T_{C1})/(T_1 - T_{C10}))}$$

(14)

- The abovementioned heat exchangers (i.e., intercooler, regenerator, high temperature heat addition, reheater, and low temperature heat rejection) are of counter-flow types, and their effectiveness can be calculated (e.g., [35]) as follows:

$$\varepsilon_i = \frac{1 - \exp[-\text{NTU}(1 - C^*)]}{1 - C^* \exp[-\text{NTU}(1 - C^*)]},$$

$$i = \text{int, reg, add, reh, rej}$$

(15)

where C^* is the ratio $(C^* = \min(C_{\text{cold}}, C_{\text{hot}})/\max(C_{\text{cold}}, C_{\text{hot}}))$ and NTU is the number of transfer unit $(\text{NTU} = UA/\min(C_{\text{cold}}, C_{\text{hot}}))$.

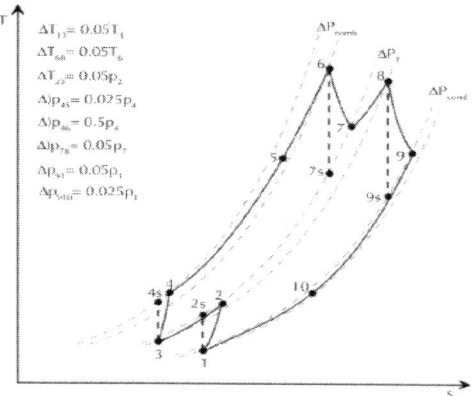

Figure 2: *T-s* diagram of realistic Brayton cycle.

Cycle Performance Parameters

Heat added to the system along processes 5 to 6 and 7 to 8 and heat rejected from system through processes 10 to 1 and 2 to 3 are given in terms of enthalpy as follows:

$$\dot{Q}_{add} = \dot{m}\left[(h_6 - h_5) + (h_8 - h_7)\right]$$

(16)

$$\dot{Q}_{rej} = \dot{m}\left[(h_{10} - h_1) + (h_2 - h_3)\right]$$

(17)

where $h_6 > h_8$ because $T_6 > T_8$ (assuming that $\Delta T_{86} = T_6 - T_8 = 0.05$ T_6) and also $h_3 > h_1$ because $T_3 > T_1$ (assuming that $\Delta T_{13} = T_3 - T_1 = 0.05$ T_1).

The power produced by both LP and HP turbines (W_t) is partially consumed by both LP and HP compressors (W_c), and the remaining power is the net power (W_{net}) as follows:

$$\dot{W}_t = \dot{m}\left[(h_6 - h_7) + (h_8 - h_9)\right]$$

(18)

$$\dot{W_c} = \dot{m}\left[(h_2 - h_1) + (h_4 - h_3)\right] \tag{19}$$

$$\dot{W}_{net} = \dot{W_t} - \dot{W_c} \tag{20}$$

The back work ratio (BWR) and first and second law thermal efficiencies (η_I, η_{II}) of the cycle are as follows:

$$BWR = \frac{\dot{W_c}}{\dot{W_t}} \tag{21}$$

$$\eta_I = \frac{\dot{W}_{net}}{\dot{Q}_{add}} = 1 - \frac{\dot{Q}_{rej}}{\dot{Q}_{add}} \tag{22}$$

$$\eta_{II} = \frac{\dot{W}_{net}}{\dot{W}_{net,rev}} = \frac{\dot{W}_{net}}{\dot{W}_{net} + \dot{X}_{dest}} \tag{23}$$

where X_{dest} is the rate of exergy destruction defined, with respect to the dead state temperature T_0 as follows:

$$\dot{X}_{dest} = T_0 (\Delta \dot{S}_{12} + \Delta \dot{S}_{23-C2C3} + \Delta \dot{S}_{34} + \Delta \dot{S}_{45-910}$$
$$+ \Delta \dot{S}_{56-H5H6} + \Delta \dot{S}_{67} + \Delta \dot{S}_{78-H7H8} + \Delta \dot{S}_{89}$$
$$+ \Delta \dot{S}_{101-C10C1}) \tag{24}$$

where the above entropy changes are calculated according to [34], taking into consideration the temperature-dependent specific heats.

For the sake of ecological performance of the cycle, and its effect on environment, the following ecological coefficient of performance (ECOP), as was previously introduced by [36,37], is defined as the power output per unit loss rate of availability as follows:

$$ECOP = \frac{\dot{W}_{net}}{\dot{X}_{loss}} \tag{25}$$

Solution Procedure

The above set of equations represents complete thermodynamic modeling of the cycle, whose solution gives the cycle performance as dependent on its controlling parameters. Following conventionally reported methods of varying one or two of the controlling parameters at a time while keeping the rest of the constants will produce some specific performance results that will be valid only for those specific variation cases and cannot be of general practical applicability. Besides, these conventional solution methods can result in localized optimized performance values that are dependent on the specific values selected for the controlling parameters. To overcome these two issues (i.e., the generalization of the study and the global optimization), the present paper has adapted the Monte Carlo methodology (MCM) that concurrently searches the variation ranges of all controlling parameters at the same time to optimize the cycle performance over the whole domain of variations of all cycle controlling parameters.

MCM Optimization Technique

The procedure of utilizing the MCM technique can be summarized as follows: (1) selection of the design and operating controlling parameters of the cycle, (2) selection of their practical variation ranges, (3) selection of the performance parameters sought to be optimized, (4) setting an acceptance-rejection criterion for the resulting performance values, (5) random selection of one complete set of values of all the controlling parameters within their variation ranges, (6) solution of the model equations (i.e., Equations 1, 2, 3, 4, 5, 6, 7 8, 9, 10, 11, 12, 13, 14, 15, 16, 17, 18, 19, 20, 21, 22, 23, 24, 25) for cycle performance to get a complete set of results based on the randomly selected set of controlling parameters, (7) applying the acceptance-rejection criterion to discard the unwanted performance values and to record the rest, and (8) repeating the above steps for another random selection of another complete set of values for the controlling parameters. The above eight steps are discussed as follows:

The design and operating parameters are as follows: inlet temperature and pressure to LP compressor $T1$, P_1; maximum temperature $T6$ entering HP turbine; pressure ratios r_{p12} and r_{p34} of LP and HP compressors; pressure ratio, r_{p67} of HP turbine; compressors and turbine efficiencies η_{c12}, η_{c34}, η_{t67}, and η_{t89}; and effectiveness of intercooler, regenerator, heat addition, reheater, and heat rejection ε_{int}, ε_{reg}, ε_{add}, ε_{reh}, and ε_{rej}, respectively. To reflect the commonly used realistic literature values, survey ranges of the controlling parameters are selected as shown in Table 1. The acceptable-rejection criteria used to disregard non-realistic performance values includes many conditional terms such as (and not limited to) follows: rejection of calculations based on violation of the second law of thermodynamics, exergy loss is negative, negative values of cycle efficiency, negative values of network, efficiencies higher than unity, unrealistic ratio of specific volumes of the two compressors, unrealistic ratio of the works of the two turbines,...etc. Based on random independent selections of values of the controlling parameters within their variation ranges, 5,000 complete calculation sets of cycle performance evaluation have been executed. Applying the acceptable-rejection criterion to these 5,000 sets of calculations has resulted on accepting only 345. The surveyed ranges of values of the controlling parameters given in the first column of Table 1 have been readjusted into acceptable ranges as shown in the second column in the same table. The results are discussed below.

Table 1: Surveyed ranges and accepted ranges of the cycle controlling parameters

Cycle controlling parameter	Surveyed range	Accepted range by MCM
T_1 entering LP compressor, (K)	300 to 450	300 to 448
P_1 entering to LP compressor, (kPa)	100 to 500	100 to 499
T_6 entering HP turbine, (K)	800 to 1500	973 to 1,483
LP compressor pressure ratios r_{p12}	1.2 to 5.4	1.281 to 5.393
HP compressor pressure ratios r_{p34}	1.2 to 5.4	1.359 to 5.393
HP turbine pressure ratio r_{p67}	1.2 to 5.4	1.353 to 5.397
η_{c12} of LP compressor	0.7 to 0.9	0.7024 to 0.8995
η_{c34} of LP compressor	0.7 to 0.9	0.7002 to 0.9000
η_{t67} of HP turbine	0.7 to 0.9	0.7000 to 0.8998
η_{t89} of LP turbine	0.7 to 0.9	0.7002 to 0.8994
ε_{int} of intercooler	0.7 to 0.95	0.7000 to 0.9500
ε_{reg} of regenerator	0.7 to 0.95	0.7012 to 0.9496
ε_{reh} of reheater	0.7 to 0.95	0.7010 to 0.9490
ε_{bur} of high temperature heat addition	0.7 to 0.95	0.7010 to 0.9496
ε_{rej} of low temperature heat rejection	0.7 to 0.95	0.7003 to 0.9495

Al-Sood *et al.*

Al-Sood *et al. International Journal of Energy and Environmental Engineering* 2013 4:37, doi:10.1186/2251-6832-4-37

RESULTS AND DISCUSSIONS

Validation of the Present Model

The operating parameters of the present model have been modified to agree with those employed in the theoretical model of Tyagi et al. [13]. Variations of first law efficiency and dimensionless power output with the low pressure turbine exit temperature for the present model and its comparison of Tyagi et al. [13] are illustrated in Figures 3 and 4, respectively. Comparisons show slight deviations that could be attributed to the pressure drop employed in the present model and neglected in Tyagi et al. model.

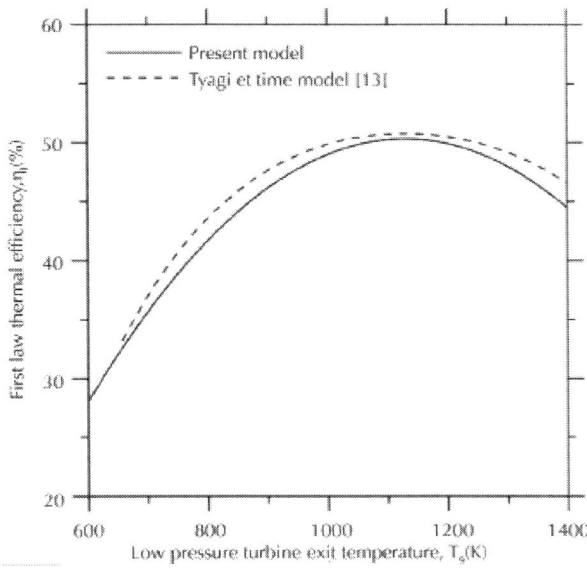

Figure 3: Variation of first law efficiency with low pressure turbine inlet temperature.

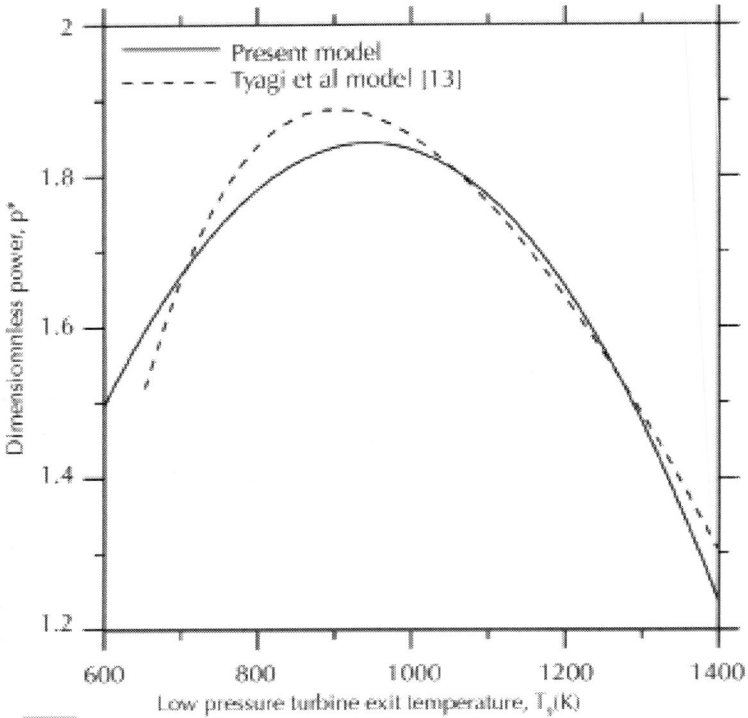

Figure 4: Variation of dimensionless power with low pressure turbine inlet temperature.

Sensitivity Analysis

The dependency of the performance parameters on the controlling parameters are displayed below as dependents, η_{I}, η_{II}, BWR, ECOP, x_{loss}, W_{net}, and q_{add}, and independents, T_{1}, T_{6}, P_{4}, and conductance of the whole cycle (i.e., summation of heat transfer coefficient-area product for all heat transfer units) UA. The shown figures display the 345 accepted results plotted as scattered points to relate the performance parameters to the controlling parameters. Each point on any of these figures represents a complete set of accepted cycle calculation, with controlling parameter values that lie within their variation ranges. Optimal performance values and the required operating parameter ranges are discussed in the following sections.

Sensitivity of Cycle Performance to Lowest Cycle Temperature T_1

Figure 5a,b,c,d,e,f,g shows the dependency of cycle performance on T_1 at values of other controlling parameters that lie within their variation ranges in Table 1. Values of η_1 in Figure 5a is very sensitive to T_1 where it exhibits a steep decrease with T_1, where its optimum values >38% that lie in the T_1 range of about 301 to 389 K, regardless of the values of all other controlling parameters. This signifies that, outside this T_1 range, no modifications of other design or operating parameters can enhance the values of η_1 beyond 38%. As expected, the lower the value of T_1, the higher is the value of η_1, with its optimum value decrease from about 48% to about 38% within this 301 to 389 K range. Figure 5b,c,d,e,f,g shows that the abovementioned range of T_1 results in optimum η_{11} in the range 33% to 66%, optimum ECOP within 1.56 to 1.92, optimum x_{loss} within 0.093 to 0.525 MJ/kg, optimum BWR within 0.473 to 0.6, optimum w_{net} within 0.178 to 0.341 MJ/kg, and optimum q_{add} within 0.422 to 0.835 MJ/kg. In these figures, respectively, ranges of T_1 are 301 to 412 K for $\eta_{11} \geq 60\%$ and ECOP ≥ 1.65, 302 to 417 K for $x_{loss} \leq 0.150$ MJ/kg, 301 to 315 K for BWR ≤ 0.525, 301 to 369 K for $w_{net} \geq 0.300$ MJ/kg, and 302 to 352 K for $q_{add} \leq 0.47$ MJ/kg. Effects of these ranges on other performance parameters are listed in Table 2. Although the performance values of these parameters suffer some deteriorations outside the abovementioned ranges of T_1, yet, and except for BWR, their sensitivity towards T_1 is not too critical. Variations of BWR show steep losses with the values of T_1.

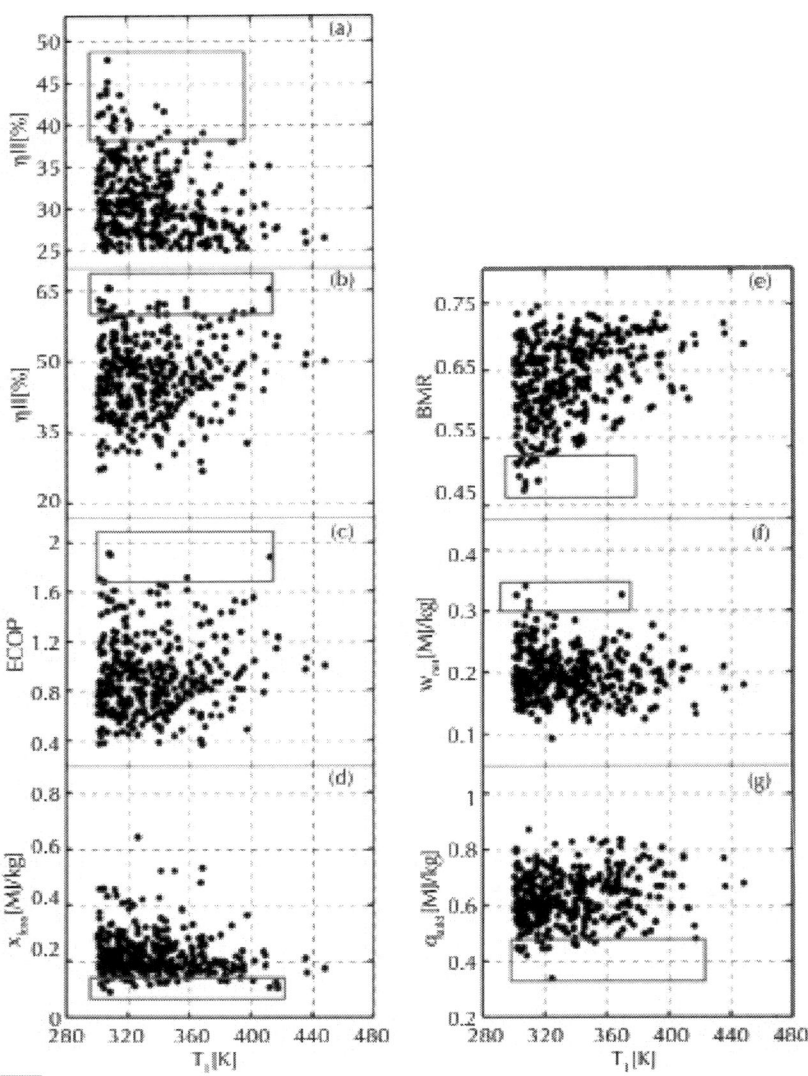

Figure 5: a-g Cycle performance parameters $\eta 1$, ηII, ECOP, $x$loss, BWR, $w$net, and $q$add versus inlet air temperature $T1$.

Table 2: Optimized performance parameters with their respected ranges of controlling parameters

	$\eta I \geq$ 38%	$\eta II \geq$ 60%	ECOP ≥ 1.65	xloss ≤ 0.150 [MJ/kg]	BWR ≤ 0.525	wnet ≥ 0.300 [MJ/kg]	qadd ≤ 0.470 [MJ/kg]
Design and performance parameters							
T_1, K	301 to 389	301 to 412	301 to 412	302 to 417	301 to 315	301 to 369	302 to 352
T_6, K	1,220 to 1,480	1,200 to 1,480	1,220 to 1,480	1,000 to 1,420	1,220 to 1,480	1,340 to 1,480	1,000 to 1,360
P_4, kPa	750 to 7,570	750 to 7,570	864 to 4,490	750 to 4,490	864 to 4,030	1,440 to 7,570	864 to 2,830
UA, kW/K	13.6 to 37.0	16.8 to 37.0	20.7 to 34.7	14.8 to 37	14.2 to 34.7	16.6 to 29.6	13.8 to 34.7
Optimum ranges of performance parameters that are achieved by ranges of operating parameters shown above							
η_I	38 to 48	32 to 48	35 to 48	25 to 44	35 to 48	36 to 48	27 to 44
η_{II}	33 to 66	60 to 66	63 to 66	45 to 66	39 to 66	48 to 66	37 to 66
ECOP	1.54 to 1.92	1.56 to 1.92	1.69 to 1.92	1.01 to 1.91	0.79 to 1.92	1.53 to 1.92	0.98 to 1.91
x_{loss}, MJ/kg	0.093 to 0.525	0.093 to 0.199	0.093 to 0.191	0.093 to 0.150	0.093 to 0.356	0.177 to 0.337	0.093 to 0.248
BWR	0.473 to 0.600	0.473 to 0.640	0.479 to 0.608	0.487 to 0.708	0.473 to 0.523	0.479 to 0.577	0.511 to 0.690
w_{net}, MJ/kg	0.178 to 0.341	0.178 to 0.341	0.178 to 0.341	0.093 to 0.246	0.178 to 0.341	0.305 to 0.341	0.093 to 0.197
q_{add}, MJ/kg	0.422 to 0.835	0.422 to 0.808	0.422 to 0.794	0.340 to 0.674	0.422 to 0.803	0.711 to 0.874	0.340 to 0.469

Al-Sood et al.

Al-Sood et al. International Journal of Energy and Environmental Engineering 2013 4:37, doi:10.1186/2251-6832-4-37

Sensitivity of Cycle Performance to Maximum Cycle Temperature T_6

Compared to the almost unified range of T_1 discussed above that produce optimum values of all the performance parameters, Figure 6a,b,c,d,e,f,g shows that T_6 has drastically changed ranges depending on which performance parameter is to be optimized. Same optimum values of η_I, η_{II}, ECOP, x_{loss}, BWR, w_{net}, and q_{add} mentioned previously and listed in Table 2 require T_6 to be in the ranges 1,220 to 1,480 K, 1,200 to 1,480 K, 1,220 to 1,480 K, 1,000 to 1,420 K, 1,220 to 1,480 K, 1,340 to 1,460 K, and 1,000 to 1,380 K, respectively. These values are generally expected since the higher is the T_6, the better η_I, η_{II}, ECOP, and w_{net}. Optimum values of the other two performance parameters, i.e., x_{loss} and q_{add}, necessitate that T_6 must be low in the range 1,000 to 1,380 K. In regards to sensitivity, and except for x_{loss} and q_{add} which are less sensitive to T_6, all other performance parameters exhibit great sensitivity to T_6, where their values greatly deteriorate outside the abovementioned optimum ranges of T_6. The wide ranges of T_6 mentioned above for optimum performance are in favor of the practical application of the cycle, which signifies that the cycle can accommodate any minor deterioration of the cycle components that are dependent on this high temperature. It is worthy to mention here that the material selection of the cycle components that are exposed to this high cycle temperature will put further restrictions and some adjustments to make these ranges practically appropriate.

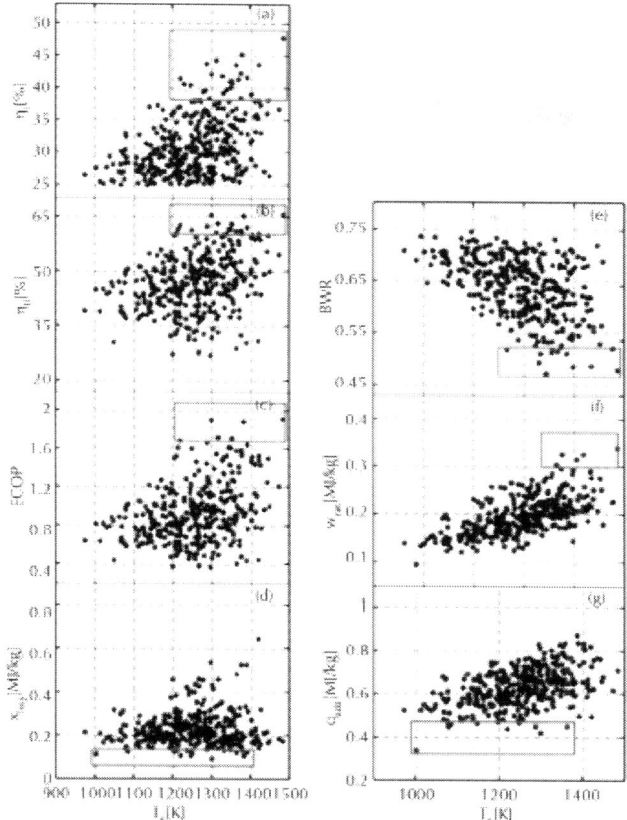

Figure 6: a-g Cycle performance parameters $\eta1$, ηII, ECOP, xloss, BWR, wnet, and qadd versus maximum cycle temperature $T6$.

Sensitivity of Cycle Performance to Maximum Cycle Pressure P_4

The effects of maximum pressure P_4 on optimum performance are shown in Figure 7a,b,c,d,e,f,g. Optimum values of η_I, η_{II}, ECOP, x_{loss}, BWR, w_{net}, and q_{add} require P_4 to be in the ranges 0.75 to 7.57 MPa, 0.75 to 7.57 MPa, 0.864 to 4.49 MPa, 0.75 to 4.49 MPa, 0.864 to 4.03 MPa, 1.44 to 7.57 MPa, and 0.864 to 2.830 MPa, respectively. In contrast to T_6, the lower the P_4, the better the cycle is. Optimum exergy loss and heat added to the cycle necessitate that P_4 must be low (i.e., in the range of 0.864 to 2.830 MPa, Figure 7d,g) to result in less losses and less amount of heat added. Although all performance parameters show different degrees of sensitivity to the value of P_4, where they show some deterioration outside the abovementioned optimum ranges of the pressure, yet w_{net} has the least sensitivity. Although pressure values up to 12 MPa have been used in the MCM, the maximum value that results in optimum value of any of the performance parameters never exceeds 7.57 MPa, which is greatly in favor of practical applications of the cycle. Again, material selections of components that are exposed to this high pressure may have some limitations imposed by their stress requirement and pumping losses.

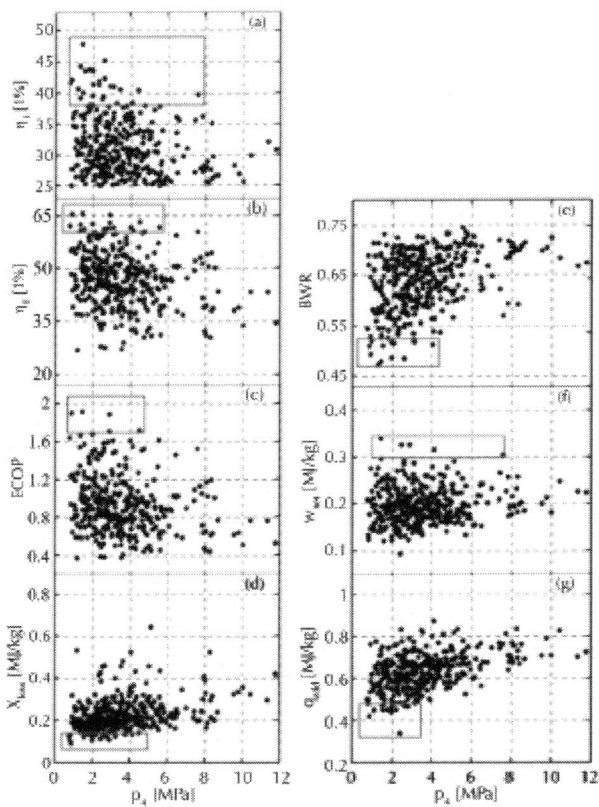

Figure 7: a-g Cycle performance parameters η1, ηII, ECOP, xloss, BWR, wnet, and qadd versus maximum cycle pressure P4.

Sensitivity of Cycle Performance to Heat Exchanger's Conductance UA

The heat exchanger's conductance, defined as the product of overall heat transfer coefficient and surface area of the heat exchanger $(UA = \dot{Q}_{add}/\Delta T_m)$, is considered an important operating/design parameter that is to be optimized based on the first law of thermodynamics and cost analysis. The selection of an optimum range for UA of heat exchangers is illustrated in Figure 8a,b,c,d. Optimum values of η_{I}, η_{II}, ECOP, x_{loss}, BWR, w_{net}, and q_{add} require UA to be in the ranges 13.6 to 37 kW/K, 16.8 to 37 kW/K, 20.7 to 34.7 kW/K, 14.8 to 37 kW/K, 14.2 to 34.7 kW/K, 16.6 to 29.6 kW/K, and 13.6 to 34.7 kW/K, respectively. All optimum cycle performance parameters require almost the same wide range of UA which is considered in favor of the cycle practical use. Although among the performance parameters, only ECOP and w_{net} show higher sensitivity with UA, where their values deteriorate very much outside their respective optimum ranges of UA, yet the non-sensitivity of the other performance parameters with UA is considered another positive point from a practical point of view.

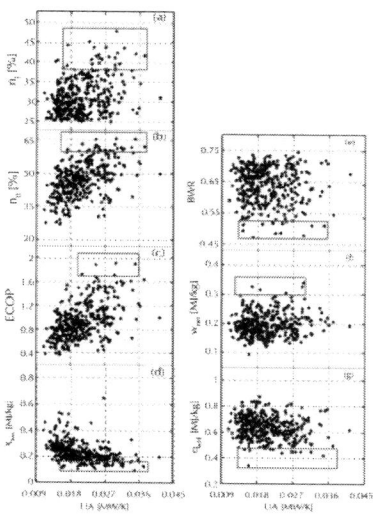

Figure 8: a-g Cycle performance parameters $\eta 1$, ηII, ECOP, $x$loss, BWR, $w$net and $q$add versus total heat transfer coefficient-area product.

Unified Operating Ranges for Simultaneous Optimum Performance

Table 3 shows the unified ranges of the operating parameters that give simultaneous optimum performance (maximum η_I, η_{II}, ECOP, w_{net}, x_{loss}, BWR, q_{add}) for the cycle. Inspection of the ranges discussed in the above sections leads to the conclusion that there are some unified ranges of the operating parameters that simultaneously optimize all the performance parameters. These ranges are as follows: T_1 (302 to 315 K), T_6 (1,340 to 1,360 K), P_4 (1.440 to 2.830 MPa), and UA (20.7 to 29.6 kW/K). Although the unified ranges for both T_1 and T_6 are very narrow, which might represent some restrictions, the good design of the components of the cycle can cope with these narrow ranges.

Table 3: Simultaneously optimum operating design parameters to achieve optimum performance parameters of an irreversible gas turbine Brayton cycle

Design parameters	Optimum range
Compressor inlet air temperature, T_1, K	302 to 315
Maximum cycle temperature, T_6, K	1,340 to 1,360
Maximum cycle pressure P_4, kPa	1,440 to 2,830
Heat exchanger conductance, UA, kW/K	20.7 to 29.6

Al-Sood et al.

Al-Sood et al. International Journal of Energy and Environmental Engineering 2013 4:37, doi:10.1186/2251-6832-4-37

Generalized Optimal Performance Equations

From the MCM results and their representative figures discussed above, least-square fitting of the data of each performance parameter with each operating parameter, that only lie on the optimum envelop (i.e., maximum or minimum), gives the following equations together with their regression coefficients R^2 and the respective ranges of its application. These equations are displayed in Table 4. Figure 9 exemplifies one set of the fitted equations (i.e., optimal η_I, η_{II}, and

ECOP versus T_1). Effects of T_1, T_6, P_4, and UA on each performance parameter is shown in Table 4.

Table 4: Coefficient of least-square fitting of the data of each performance parameter with operating parameter

		Polynomial coefficients $y = \sum_{i=0}^{4} a_i x^i$					
		a0	a1	a2	R2	Range of y	Range of x
Effect of T1 (K)	$\eta 1$ (%)	38.70743	0.14107	-3.71733×10^{-5}	0.977	47.88% to 26.5%	307 to 448 K
	ηII (%)	-10.80760	0.49830	-8.11052×10^{-5}	0.99	56.76% to 50.21%	307 to 448 K
	ECOP	0.10328	0.01473	-2.85902×10^{-5}	0.98	1.92 to 1.01	307 to 448 K
	xloss (kJ/kg)	38.7074	0.14107	-0.0003	0.97	47.88 to 26.5 kJ/kg	307.448 K
	BWR	-3.98656	0.02608	-0.0000373	0.99	0.5234 to 0.5758	311 to 345 K
	wnet (kJ/kg)	4,098.474	22.73832	0.033916	0.95	285.2 to 326.5 kJ/kg	301 to 3,689 K
	qin (kJ/kg)	-768.6663	6.51454	-0.008414	0.85	440.1 to 500.3 kJ/kg	304 to 417 K
Effect of T6 (K)	$\eta 1$ (%)	19.9733	0.11057	5.18151×10^{-6}	0.88	40.64% to 47.88%	1,219 to 1,483 K
	ηII (%)	-135.8492	0.286828	-1.01925×10^{-4}	0.91	61.74% to 65.76%	1,213 to 1,483 K
	ECOP	-2.06871	4.66995×10^{-3}	-1.33033×10^{-6}	0.9	1.92 to 1.61	1,213 to 1,483 K
	xloss (kJ/kg)	2.17010×10^{3}	-3.31852	1.33394×10^{-3}	0.92	107.7 to 177.4 kJ/kg	1,021 to 1,483 K
	BWR	4.54487	-6.027229×10^{-3}	2.23556×10^{-6}	0.91	0.485 to 0.521	1,219 to 1,472 K
	wnet (kJ/kg)	-1.35172×10^{3}	2.1247	-6.63125×10^{-4}	0.97	304.5 to 340.6 kJ/kg	1,338 to 1,483 K
	qin (kJ/kg)	593.0112	-0.69715	4.44143×10^{-4}	0.93	340 to 499.2 kJ/kg	1,002 to 1,421 K

Effect of $P4$ (MPa)	η1 (%)	39.4442	3.78362×10^{-3}	-5.15619×10^{-7}	0.95	45.25% to 35.22%	0.75 to 8.32 MPa
	ηII (%)	-4.93274×10^{-8}	-4.9076×10^{-4}	66.37495	0.95	65.76% to 60.49%	0.86 to 7.57 MPa
	ECOP	1.88093	5.12392×10^{-5}	-1.81698×10^{-8}	0.99	1.92 to 1.62	0.86 to 5.53 MPa
	xloss (kJ/kg)	-4.04347×10^{-7}	0.015887	79.08315	0.99	163.2 to 93.3 kJ/kg	0.85 to 6.43 MPa
	BWR	8.93573×10^{-9}	-3.58163×10^{-5}	0.51752	0.96	0.537 to 0.479	0.923 to 4.41 MPa
	wnet (kJ/kg)	9.38396×10^{-7}	-0.0142	358.3297	0.99	340.5 to 304.5 kJ/kg	1.44 to 7.57 MPa
	qin (kJ/kg)	2.47949×10^{-6}	0.01457	405.67103	0.99	502.5 to 421.5 kJ/kg	0.86 to 3.94 MPa
Effect UA (kW/K)	η1 (%)	16.2101	2.39670	-0.04650	0.86	47.88% to 40.14%	13.61 to 37.04 kW/K
	ηII (%)	32.70171	2.35327	-0.04146	0.9	65.76% to 60.78%	16.79 to 37.4 kW/K
	ECOP	-0.03198	0.13002	-2.14243×10^{-3}	0.98	1.92 to 1.55	16.79 to 34 67 kW/K
	xloss (kJ/kg)	0.10875	-5.16757	168.692	0.99	126.4 to 107.7 kJ/kg	15.9 to 37.04 kW/K
	BWR	4.35305×10^{-4}	-0.0201	0.68964	0.98	0.545 to 0.473	14.22 to 37.04 kW/K
	wnet (kJ/kg)	-0.23225	12.10661	178.35128	0.73	340.6 to 30.1 kJ/kg	13.6 to 29.56 kW/K
	qin (kJ/kg)	0.33363	-16.06035	6.33.8598	0.94	399.2 to 440.1 kJ/kg	13.31 to 26.78 kW/K

Al-Sood et al.

Al-Sood et al. International Journal of Energy and Environmental Engineering 2013 4:37, doi:10.1186/2251-6832-4-37

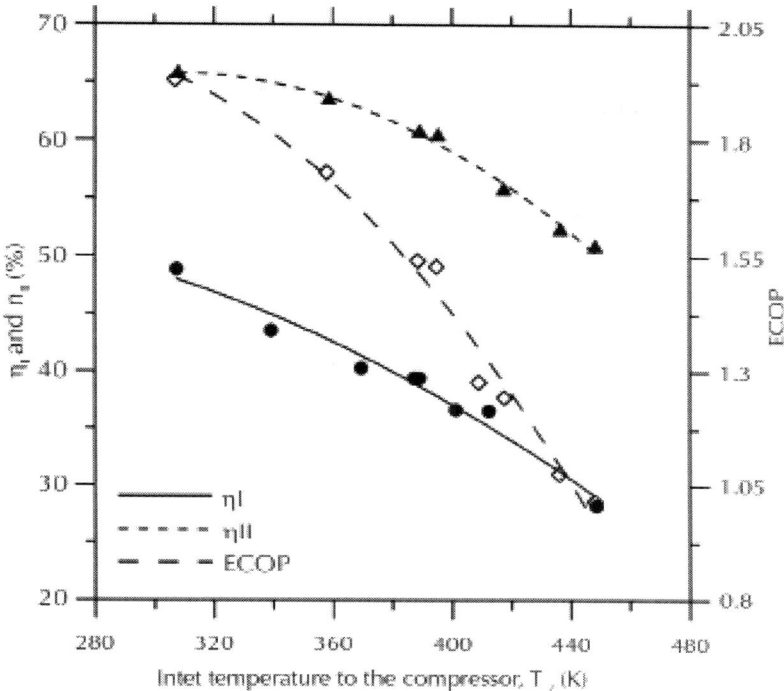

Figure 9: Optimum MCM results of $\eta 1$, ηII, and ECOP and their fitted equations w.r.t. T1. With ranges of other operating parameter values as in Table 1.

The set of equations displayed in Table 4 can form a good basis for designing an optimal cycle, where the effect of each of the operating parameters on each of the performance parameters has been exclusively demonstrated in this mathematical form along with the applicable ranges of these two parameters regardless of the values of the other parameters. It is worthy to mention that the above equations are the result of a survey that concurrently covers all the practical ranges of the operating parameters, which can be easily understood to be the global optimal representation of the performance of the cycle. Also, the results discussed above are generally applicable to the cycle and are not restricted to some specific values of operating parameters or conditions of operation.

CONCLUSIONS

The present study has developed a general mathematical model to specify the performance as dependent on design and operating parameters of an irreversible gas turbine Brayton cycle incorporating two-stage compressor, two-stage gas turbine, intercooler, reheater, and regenerator with irreversibilities due to finite heat transfer rates and pressure drops. Ranges of operating parameters resulting in optimum performance (i.e., $\eta_{\text{I}} \geq 38\%$, $\eta_{\text{II}} \geq 60\%$, ECOP ≥ 1.65, $x_{\text{loss}} \leq 0.150$ MJ/kg, BWR ≤ 0.525, $w_{\text{net}} \geq 0.300$ MJ/kg, and $q_{\text{add}} \leq 0.470$ MJ/kg) are determined and discussed using the Monte Carlo method. These operating ranges are as follows: minimum cycle temperature ranges between 302 and 315 K, maximum cycle temperature ranges between 1,320 and 1360 K, maximum cycle pressure ranges between 1.449 and 2.830 MPa, and conductance of the heat exchanger ranges between 20.7 and 29.6 kW/K. The exclusive effect of each of the operating parameters on each of the performance parameters is mathematically given in a general sense that is applicable regardless of the values of the rest of the operating parameters and under any condition of operation of the cycle.

AUTHORS' CONTRIBUTIONS

MMAAS conceived the concept and procedures of the present work, developed the model, carried out the analysis of the results, and wrote the manuscript. KKM checked the equations and analysis and reviewed the manuscript. YMAR developed the model, carried out the computations, and reviewed the manuscript. All authors read and approved the final manuscript.

ACKNOWLEDGMENT

This work has been fully supported by the Assiut University and the Mechanical Engineering Department.

REFERENCES

1. Kaushik, SC, Tyagi, SK: Finite time thermodynamic analysis of a nonisentropic regenerative Brayton heat engine. Int. J. Sol. Energy. 22, 141–151 (2002).

2. Rahman, MM, Ibrahim, TK, Taib, MY, Noor, MM, Bakar, RA: Thermal analysis of open-cycle regenerator gas-turbine power-plant. World Academy of Science Eng Technol. 44, 1307–13012 (2010)

3. Vecchiarelli, J, Kawall, JG, Wallace, JS: Analysis of a concept for increasing the efficiency of a Brayton cycle via isothermal heat addition. Int. J. Energy Res. 2, 113–127 (1997)

4. Göktun, S, Yavuz, H: Thermal efficiency of a regenerative Brayton cycle with isothermal heat addition. Energy Convers. Manage. 40, 1259–1266 (1999).

5. Erbay, LB, Göktun, S, Yavuz, H: Optimal design of the regenerative gas turbine engine with isothermal heat addition. Appl. Energy. 68, 249–269 (2001).

6. Kaushik, SC, Tyagi, SK, Singhal, MK: Parametric study of an irreversible regenerative Brayton heat engine with isothermal heat addition. Energy Convers. Manage. 44, 2013–2025 (2003).

7. Cheng, CY, Chen, CK: Maximum power of an endoreversible intercooled Brayton cycle. Int. J. Energy Res. 24, 485–49 (2000).

8. Canie`re, H, Willockx, A, Dick, E, Paepe, MD: Raising cycle efficiency by intercooling in air-cooled gas turbines. Appl Therm Eng. 26, 1780–1787 (2006).

9. Negridi, MG, Gambini, M, Peretto, A: Reheat and regenerative gas turbine for feed water repowering of steam power plant, Houston: ASME Turbo Expo (1995)

10. Khaliq, A, Kaushik, SC: Thermodynamic performance evaluation of combustion gas turbine cogeneration system with reheat. Appl Therm Eng. 24, 1785–1795 (2004).

11. Hernández, C, Roco, JMM, Medina, A: Power and efficiency in a regenerative gas-turbine with multiple reheating and intercooling stages. J. Phys. D: Appl. Phys. 29, 1462–1468 (1996).

12. Sogut, OS, Ust, Y, Sahin, B: The effects of intercooling and regeneration on the thermo-ecological performance analysis

of an irreversible-closed Brayton heat engine with variable temperature thermal reservoirs. J. Phys. D: Appl. Phys. 39, 4713–4721 (2006).

13. Tyagi, SK, Chen, GM, Wang, Q, Kaushik, SC: Thermodynamic analysis and parametric study of an irreversible regenerative-intercooled-reheat Brayton cycle. Int J Therm Sci. 40, 829–840 (2006)

14. Wang, W, Chen, L, Sun, F, Wu, C: Performance analysis of an irreversible variable temperature heat reservoir closed intercooled regenerated Brayton cycle. Energy Convers. Manage. 44, 2713–2732 (2003).

15. Sánchez-Orgaz, S, Medina, A, Hernández, AC: Thermodynamic model and optimization of a multi-step irreversible Brayton cycle. Int J Therm Sci. 51, 2134–2143 (2010)

16. Herrera, A, Sandoval, JA, Rosillo, ME: Power and entropy generation of an extended irreversible Brayton cycle: optimal parameters and performance. J. Phys. D: Appl. Phys.39, 3414–3424 (2006).

17. Landsberg, PT, Leff, HS: Thermodynamic cycles with nearly universal maximum-work efficiencies. J. Phys. A: Mathematical and General. 22, 4019–4026 (1989). Aragón-González, G, Canales-Palma, A, León-Galicia, A: Maximum irreversible work and efficiency in power cycles. J. Phys. D: Appl. Phys. 33, 1403–1409 (2000)

18. Roco, JMM, Velasco, S, Medina, A, Hemandez, AC: Optimum performance of a regenerative Brayton thermal cycle. J Appl Phys. 82, 2735–2741 (1997).

19. Wu, C, Chen, L, Sun, F: Performance of a regenerative Brayton heat engine. Energy. 21, 71–76 (1996).

20. Ibrahim, TK, Rahman, MM: Effects of operation conditions on performance of a gas turbine power plant. National Conference in Mechanical Engineering Research and Postgraduate Studies (2nd NCMER 2010), pp. 135–144. Kuantan: Faculty of Mechanical Engineering, UMP Pekan (2010)

21. Ali Mousafarash, A, Ameri, M: Exergy and exergo-economic based analysis of a gas turbine power generation system. Journal of Power Technologies. 93, 44–51 (2013)

22. Medina, A, Roco, JMM, Hernandez, AC: Regenerative gas turbines at maximum power density conditions. J. Phys. D: Appl. Phys. 29, 2802–2805 (1996).

23. Chen, L, Zheng, J, Sun, F, Wu, C: Performance comparison of an irreversible closed Brayton cycle under maximum power density and maximum power conditions. Exergy, an International Journal. 2, 345–351 (2002).

24. Al-Hadhrami, LM, Shaahid, SM, Al-Mubarak, AA: Jet impingement cooling in gas turbines for thermal efficiency and power density. In: Ernesto B (ed.) Advances in Gas Turbine Technology, pp. 191–210. New York: InTech (2011)

25. Hernández, AC, Medina, A, Roco, JMM: Power and efficiency in a regenerative gas turbine. J. Phys. D: Appl. Phys. 28, 2020–2023 (1995).

26. Li, Y, Huang, Y, Yan, X: The effects of variable specific heats of working fluid on the performance of irreversible reciprocating Brayton cycle. Advanced Materials Research.345–355, 1305–1310 (2012)

27. Stevens, T, Baelmans, M: Optimal pressure drop ratio for micro recuperators in small sized gas turbines. Appl Therm Eng. 28, 2353–2359 (2008).

28. Farzaneh-Gord, M, Deymi-Dashtebayaz, M: Effect of various inlet air cooling methods on gas turbine performance. Energy. 36, 1196–1205 (2011).

29. De Sa, A, Al Zubaidy, SA: Gas turbine at varying ambient temperature. Appl Therm Eng.31, 2735–2739 (2011).

30. Sayyaadi, H, Reza Mehrabipour, R: Efficiency enhancement of a gas turbine cycle using an optimized tubular recuperative heat exchanger. Energy. 38, 362–375 (2012).

31. Ahmadi, P, Dincer, I: Thermodynamic and exergoenvironmental analysis and multi-objective optimization of a gas turbine power plant. Appl Therm Eng. 31, 2529–2540 (2011).

32. Haseli, Y: Optimization of a regenerative Brayton cycle by maximization of a newly defined second law efficiency. Energy Convers. Manage. 68, 113–140 (2013)

33. Cengel, YA, Boles, MA: Thermodynamics: An engineering Approach, New York: McGraw Hill (2010)

34.   Shah, RK, Sekulic, DP: Fundamentals of Heat Exchanger design, New York: Wiley (2003)

35.   Ust, Y, Sahin, B, Kodal, A, Akcay, IH: Ecological coefficient of performance analysis and optimization of an irreversible regenerative-Brayton heat engine. Appl. Energy. 83, 558–572 (2006).

36.   Ust, Y, Sahin, B, Kodal, A: Performance analysis of an irreversible heat engine base on ecological coefficient of performance criterion. Brayton heat engine. Int. J. Therm. Sci. 45, 94–101 (2006).

6

Development of Semiclosed Cycle Gas Turbine for Oxy-Fuel IGCC Power Generation with CO$_2$ Capture

Takeharu Hasegawa[1]

[1]Central Research Institute of Electric Power Industry, Nagasaka, Yokosuka-Shi Kanagawa-Ken, Japan

INTRODUCTION

In response to recent changes in energy-intensive and global environmental conditions, it is urgent and crucial concern to develop the high-efficiency technologies of fossil fuel power generations. Especially, coal is one of the most important resources from the standpoint of risk avoidance in the scheme of power supply composition. Figure 1 shows the proved recoverable reserves of coal by region compared with those of the natural gas and crude oil. The world's coal reserves are twice that of each conventional oil and natural gas, distributed more evenly on a geographical basis than those for oil and natural gas, and also geopolitical risk is lower for securing the stable supply of coal resource. This figure also shows each total discoverable reserve of non-conventional resources of natural gas and crude oil as references, and each reserve corresponds to twice of the coal proved recoverable reserve. In this regard, however, total discoverable reserve of coal is estimated ten times of proved recoverable reserves, or it is corresponds to five times of that of each non-conventional resource of natural gas and crude oil. Coal is definitely the most important fossil fuel resources in the future.

Furthermore, in the 1997 when the Third Conference of Parties to the United Nations Framework Convention on Climate Change (COP3), the Kyoto protocol, which invoked mandatory CO_2 emissions reductions on countries, was adopted. CO_2 emissions per unit calorie of coal are about 1.8 times that in the case of natural gas, and then CO_2 recovery technologies are very important for thermal power plants.

On the other hand, demand of coal has increased rapidly in the recent years. Figure 2 shows annual changes of the world's coal consumption by region and the reserves-to-production ratios of coal, oil and natural gas. In the intervening quarter-century from 1985 to 2010, the coal consumption in Asia Pacific increased significantly or about 3.6 times, while world coal consumption increased 1.7 times. The increase in coal consumption in Asia Pacific is equal to one half of the world's consumption in 2010, while consumptions in other regions decrease. In just ten years, coal consumption in Asia Pacific increased double, and then the world's reserves-to-production ratio of coal decreased by half, while the reserves-to-production ratios of oil and natural gas have been maintained constant. Along with the growing world demands for fossil energy resources in recent years, international competition for development of fossil fuel fields of coal, oil and gas in the world is ever intensified.

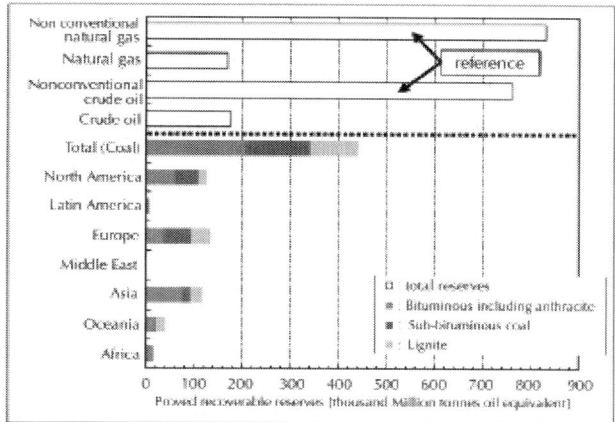

Figure 1: Proved recoverable reserves of coal by region at end 2010, compared with oil and natural gas reserves. Source of reserves data: BP statistical review of world energy 2011 [1]. Notes: Coal proved reserves expressed in tonnes oil equivalent are calculated using coal productions based on data expressed in tonnes oil equivalent and coal productions in tonnes. Nonconventional natural gas shows data not including methane hydrate reserves. Nonconventional crude oil includes oil shale and oil sand reserves.

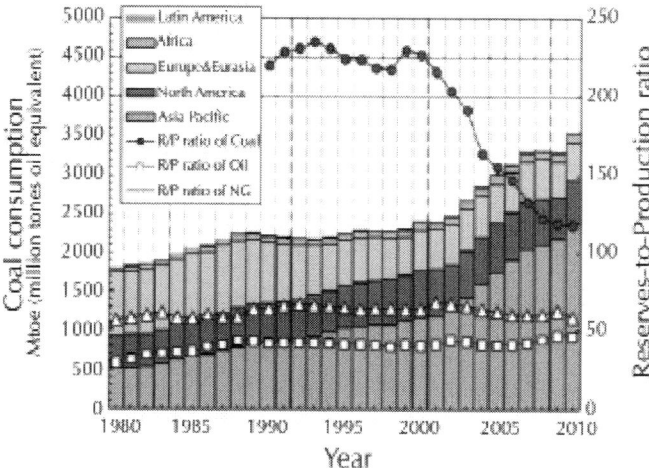

Figure 2: World coal consumption by region and proved recoverable reserves-to-production (R/P) ratio of coal, oil and natural gas (NG) at end 2010.

Note: Coal data include anthracite, bituminous, sub-bituminous, and lignite. And reserves-to-production (R/P) ratios are approximate values based on the total proved recoverable reserves of bituminous coal, anthracite, lignite and sub-bituminous coal. Sources are BP statistical review of world energy [1] and data reported for precious World Energy Council Surveys of Energy Resources [2].

With the above mentioned situations as a background, developments of high-efficiency power generation technologies and low emission technologies of CO_2 become increasingly important in the world. As one of the highly-efficient and low CO_2 emission technologies, an integrated coal gasification combined cycle (IGCC) power generation combined with CO_2 capture and storage (CCS) technologies are now drawing attention from the electric power industry. The Central Research Institute of Electric Power Industry (CRIEPI) has proposed a newly-designed oxy-fuel IGCC power generation system integrated with a combination of CO_2 recovery processing and a semiclosed cycle gas turbine [3]. This system wields the advantages of not requiring a CO_2 capture system using CO_2 absorption processing or fuel reforming preprocessing. Compared to conventional CO_2 recovery thermal power plants, oxy-fuel IGCC could simplify CO_2 recovery systems, reduce station service power, and achieve higher thermal efficiency. Currently, CRIEPI is addressing each technological development [4-9] towards the realization of highly efficient power generation with zero emissions, and with a semiclosed gas turbine system serving as one of the key technologies.

In this study, we have been researching and developing the combustion technologies in order to achieve the semiclosed cycle gas turbine for highly efficient oxy-fuel IGCC [5-6]. This paper describes technical difficulties and combustion characteristics of semiclosed gas turbine combustors, comparing developed H_2/O_2 and natural gas/O_2 fired semiclosed gas turbines in the WE-NET project [10] and a conventional natural gas fired gas turbine.

CO$_2$ RECOVERY FROM THERMAL POWER PLANT

Co$_2$ Recovery Methods for IGCCS

Along with the oxy-fuel IGCC system newly proposed in this paper, there exist four CO$_2$ recovery systems for coal-base thermal power generation. With regard to CO$_2$ recovery systems for IGCC, as shown in figure 3, the oxy-fuel IGCC system and the pre-combustion system for IGCC are under development [11-14]. In the case of an oxy-fuel IGCC power generation system with CO$_2$ capture in a semiclosed cycle oxy-fuel gas turbine, recovery of CO$_2$ is simplified, with decreasing station service power expected to produce highly efficient generation. This is because water-gas-shift reactors and physical/chemical solvents for CO$_2$ capture are not required as opposed to conventional pre-combustion systems for IGCC.

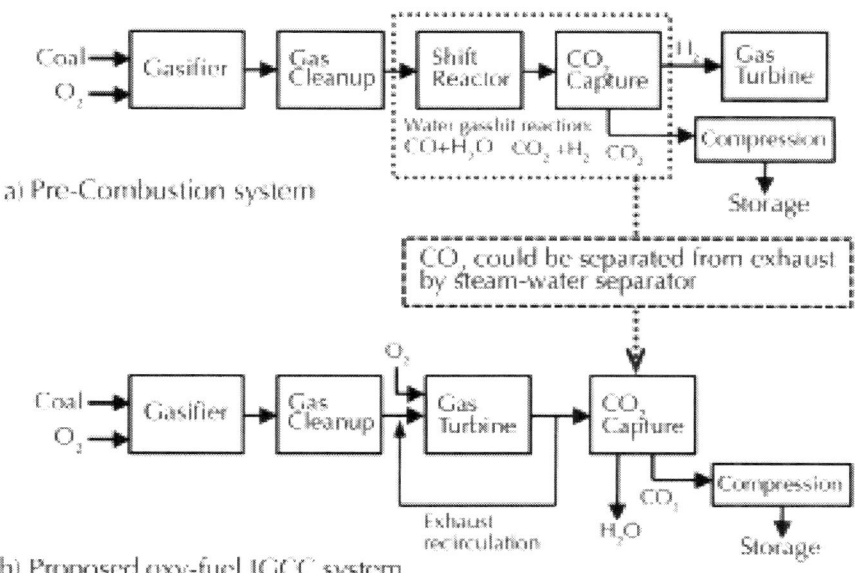

Figure 3: Comparison of CO$_2$ recovery processes for IGCCs.

Figure 4 shows the change in net plant efficiency (HHV basis) in conventional pulverized coal and IGCC power plants. In the case of 90 percent CO_2-recovery, post-combustion systems, the thermal efficiency of pulverized-coal, super critical boilers decreases to 28.4% [11] from 39.3% since a huge amount of steam is needed to regenerate absorbers, while oxy-fuel combustion systems of O_2-fired pulverized coal boilers result in only a marginal improvement in thermal efficiency of 29.3% [11]. Furthermore, in the case of a pre-combustion system using an F-class gas turbine for IGCC, thermal efficiency is expected to improve to 31.6% [11].

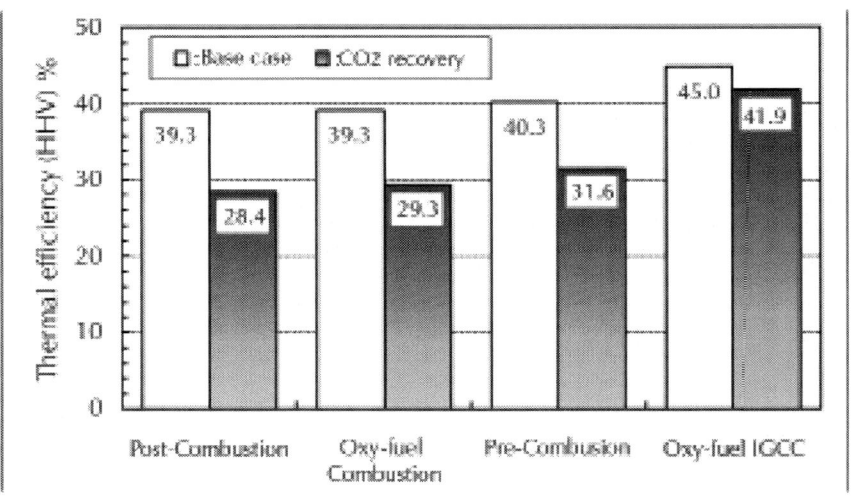

Figure 4: Thermal efficiency of coal-base power plants with and without CO_2 capture and compression. In the three conventional cases of post-, oxy-fuel and pre-combustion, currently available technologies are employed and CO_2 recovery rate is set at 90% [11]. In the case of oxy-fuel IGCC employing technologies currently under development, CO_2 recovery rate is set at 99% [4].

However, in the case of an oxy-fuel IGCC system adopting each technology currently under development, the use of an O_2-CO_2 gasifier, for example, with a hot/dry synthetic gas cleanup system and a semiclosed cycle gas turbine (turbine inlet temperature on ISO standard basis at about 1530K), is expected to produce a transmission-end thermal efficiency of 41.9% under conditions of 99% or higher CO_2 recovery.

Oxy-Fuel IGCC and Closed-cycle Gas Turbine

Figure 5 shows a schematic diagram of the oxy-fuel IGCC system and a topping semiclosed cycle gas turbine. The newly proposed oxy-fuel IGCC consists of an oxygen-CO_2 blown gasifier, a hot/dry synthetic gas cleanup system, a semiclosed cycle oxygen-fired gas turbine, and a CO_2 recovery process. This system has the following advantages;

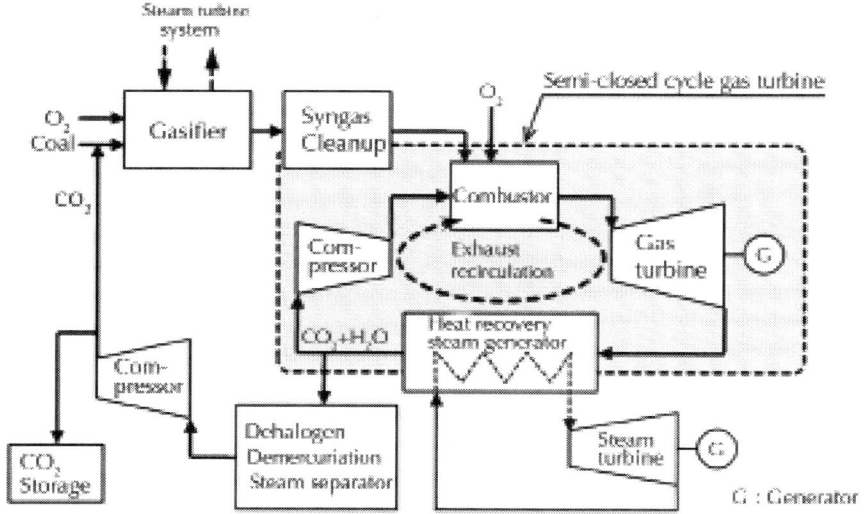

Figure 5: Schematic diagram of oxy-fuel IGCC and semiclosed cycle gas turbine [3].

- Oxygen-CO_2 blown, entrained-flow coal gasifier

Table 1 shows the rated conditions of a gasified fuel and semiclosed cycle gas turbine combustor [3],[4]. Table 2 shows characteristics of coal used in the calculation [3]. Here, we dry fed pulverized coal into an oxygen-blown entrained-flow gasifier with recycled CO_2 from flue gas, and gasified with additional oxygen. In addition, we found that O_2-CO_2 blown coal gasification enhanced gasification efficiency compared to that of current oxygen blown gasification through dry feeding of coal with N_2.Figure 6 shows the gasification characteristics of the two cases above, estimated by numerical analysis of a one-dimensional model [3].

Table 1: Rated conditions of semiclosed cycle gas turbine combustor [3],[4]

Components	Gasified fuel	Oxidizer	Dilution
CO [vol%]	66.2	0	0
H2	23.8	0	0
CH4	0.3	0	0
CO2	4.9	0	69.5
H2O	3.2	0	26.9
Ar, N2	1.5	2.5	2.7
O2	0	97.5	0.9
HHV (LHV)	11.5 MJ/m3 (11.0 MJ/m3) at 273K, 0.1MPa		
Pressure in combustor	2.2MPa		
*	0.98 (Overall equivalence ratio is 0.89)		
Dilution ratio	5.5: dilution/fuel molar ratio		
Exhaust temp.	1573K at combustor exit		
* : calculated from fuel and oxidizer without O2 concentration in dilution			

Table 2: Characteristics of coal used in calculation [3]

Inherent moisture* [wt%]	3.6
Ash content* [wt%]	9.6
Volatile matter* [wt%]	30.3
Fixed carbon* [wt%]	56.5
Ultimate analysis**	
C [wt%]	76.1
H [wt%]	5.1
O [wt%]	6.9
N [wt%]	1.7
S [wt%]	0.5

[i] - *: air-dried state, **: dry basis

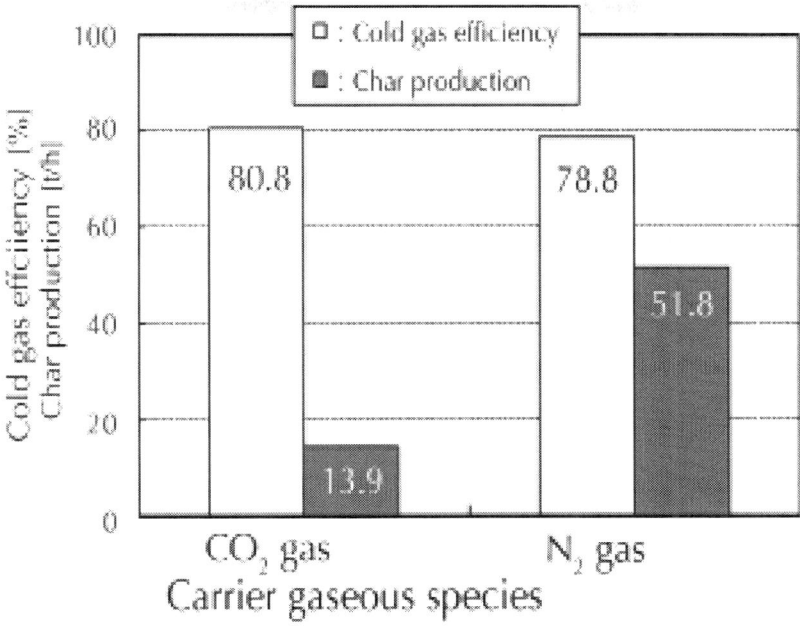

Figure 6: Influence of carrier gas conveying pulverized coal into gasifier on oxygen-blown gasification performance under conditions of coal input of 118.5t/h [3].

Table 3 shows numerical analysis conditions in gasification. Gasified fuels were calculated under conditions where an equivalence ratio in the gasification was set at 2.58 through multi-stage analyses utilizing pyrolysis, char gasification reaction and gas phase equilibrium reaction processes, and assuming a one-dimensional axial flow in the entrained-flow gasifier [3].We assumed that volatile matter contents in coal would be instantaneously pyrolyzed in the first stage, so we took 3-step reduced reactions in char gasification into account. For char gasification, we used char reaction rates based on experimental data from a pressurized drop tube furnace [15]. In the analyses, we determined the point in time when char input accorded with char production to be equilibrium. Since we assumed 100% removal rates of dust and sulfur in the synthetic gas cleanup, gasified fuels shown in Table 1 did not include sulfur, halide, and ash and metal impurities.

Table 3: Analysis method and conditions [3]

Calculation	One-dimensional model
Reaction	
1) Pyrolysis	Pyrolyzed instantaneously
Coal → CnHmOl + Char	
2) Reaction of Char	Reaction rates obtained from data in pressure drop tube furnace
C + 1/2O2 → CO	
C + CO2 → 2CO	
C + H2O → CO+H2	
3) Gas phase reaction (C,H,O)→(CH4,H2,CO,CO2,H2O,N2,O2)	Equilibrium reaction

The cold gas efficiency in Fig.6 demonstrates the ratio between chemical energy content in the product gas compared to chemical energy in fuel on a lower heating value basis. Cold gas efficiency was calculated in the following way:

$$coldgasefficiency = \frac{productgas|massflow \times heatingvalue| - additionalfuel|massflow \times heatingvalue|}{coal\ sup\ pliedtogasifier|massflow \times heatingvalue|} \times 100$$

(1)

As a result, we estimated an improvement in cold gas efficiency by 2 percent and a reduction of char particles. At the same time, we clarified the influence of CO_2 and H_2O content on char production characteristics by using a pressurized drop tube furnace [8], and we evaluated the effects of CO_2 enrichment on coal gasification performance using an actual pressurized entrained flow coal gasifier of a 3ton/day bench scale gasifier [9]. Results confirmed that CO_2 enrichment improves gasification characteristics.

- Hot/dry synthetic gas cleanup

We treated gasified fuels with a hot/dry synthetic gas cleanup system consisting of a metallic filter, a hot gas desulfurization unit and other materials, which simplified the cleanup system and reduced the power consumption for cleanup [7]. Dust removal technologies using metallic filters or ceramic ones have already been demonstrated and put to practical use in IGCC plants. So far, the Central Research Institute of Electric Power Industry has developed a halide sorbent containing $NaAlO_2$ [16], a honeycomb zinc ferrite desulfurization sorbent containing ZnFe2O4 [7], a honeycomb copper based mercury sorbent containing CuS [17], and an ammonia decomposing Ni-based catalyst supported by ZSM-5 pellets [18] and each of those elemental technologies was expected to be applied to the hot/dry synthetic gas cleanup system for current IGCCs. Figure 7 [19] shows the schematics of the demonstration plant of the dry gas purification system for the IGCC now being developed. An ammonia catalytic removal process was expected to be installed following the desulfurization unit. The process sequence of the purification system was determined by considering the operation temperature and performance of the sorbents and catalyst. Recently, the Central Research Institute of Electric Power Industry has been moving ahead on design of a new dry gas purification system for the advanced oxy-fuel IGCC by applying the purification system employing the elemental technologies developed for current IGCCs. Impurities in gasified fuels such as dust, ash contents, metal compounds, sulfur, halide, mercury and others could be reduced to an allowable level [20] for conventional gas turbines.

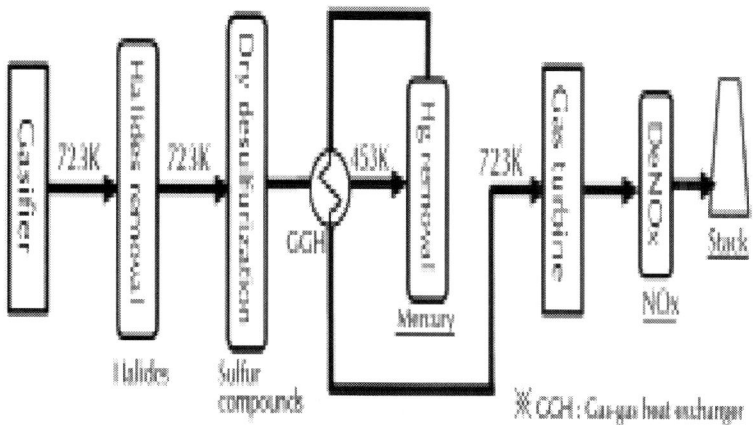

Figure 7: Schematic flow diagram of demonstration plant of dry gas purification system for current IGCCs [19].

- Semiclosed cycle gas turbine and CO_2 recovery

In a semiclosed cycle oxy-fuel gas turbine system as a topping cycle, we burned gasified fuels with pure oxygen and adjusted combustor exhaust temperature by recycling CO_2-enriched flue gas. As shown in Table 1, the rated temperature of combustor exhaust was set at 1573K (1300degC) and pressure inside the combustor at 2.2MPa [4]. After recovering exhaust heat in the HRSG, the necessary amount of flue gas was compressed and recycled to a gas turbine.

We then fed the remaining flue gas to a water scrubber of a halogen and Hg removal system and mist separator. We found that following these treatments, flue gas consisting mostly of CO_2 and H_2O became high-concentration CO_2 gas. We used some of the flue gas to feed coal to a gasifier, with the remainder compressed and sent to a storage site. It was necessary to reduce oxygen concentration in coal carrier gas to a low level in order to prevent pulverized coal from firing inappropriately.

Table 4 shows subjects and characteristics of gasified fuel/O_2 stoichiometric combustion with exhaust recirculation compared to a conventional natural gas-fired gas turbine. Unlike in the case of excess air combustion of a natural gas-fired gas turbine, the suppression of fuel oxidation under O_2-fired stoichiometric conditions with exhaust recirculation poses concerns, thereby necessitating the development of combustion promotion technology.

Table 4: Subjects of semiclosed cycle gas turbine of gasified fuel/O_2 stoichiometric combustion with exhaust recirculation

	Oxy-fuel combustion in IGCC	Conventional natural gas-fire GT
Equivalence ratio	Stoichiometric (0.98)	0.4~0.5
	Oxidation reaction is restrained and unburned fuel is emitted.	at Tex =1573K ~1773K
Dilution gas to adjust combustion temp.	Exhaust recirculation	Air
	Some exhaust is used as coal carrier gas, and then O2 concentration has to be decreased to a safe level.	

NOx emissions	Hot/dry cleanup and exhaust recirculation cause increased NOx emissions	Only thermal-NOx emissions

In the case of oxy-fuel combustion in IGCC, a little excess O_2 combustion in which apparent equivalence ratio is set at 0.98 or lower resulted in higher concentrations of residual O_2 in exhaust, restraining the usage of exhaust to feed coal into the gasifier while combustion efficiency rose. And the presence of non-condensable gases such as remaining O_2, and Ar and N_2 separated from the air resulted in increased condensation duty for the recovery of the CO_2 [21]. On the other hand, a little higher equivalence ratio over stoichiometric conditions decreased combustion efficiency. We have to accomplish higher combustion efficiency under almost stoichiometric conditions and decrease.

Furthermore, both the employment of hot/dry synthetic gas cleanup and exhaust recirculation increased fuel-NOx emissions.

Against the above backdrop, we first of all researched combustion characteristics and exhaust gas reaction characteristics in the semiclosed cycle gas turbine for oxy-fuel IGCC [5].

NUMERICAL ANALYSIS METHOD BASED ON ELEMENTARY REACTION MODELS WITH PSR AND PFR

We examined the reaction characteristics of reactant gases both in the combustor and in exhaust using numerical analysis based on the following elementary reaction kinetics. Here, we employed the reaction model proposed by Miller and Bowman [22], and confirmed by test result comparison the appropriateness of the model for non-catalytic reduction of ammonia in gasified fuel using NO [23] and an oxidation of ammonia by premixed methane flame [22].

The reaction scheme we employed was composed of 248 elementary reactions, with 50 species taken into consideration. Miller and Bowman described both a detailed scheme of the oxidation of C1 and C2 hydrocarbons under most (but not too fuel-rich) conditions, and an essential scheme for ammonia oxidation. Hasegawa et al. [23], united these two schemes and confirmed the applicable scope of a united scheme through experiments using a flow tube reactor. As an example, figure 8 shows comparative calculations results with non-catalytic denitration tests performed by Lion [24]. The analytical results precisely described a narrow reaction temperature for effective non-catalytic denitration and the behavior of NH_3 and NO constituents. Furthermore, the authors have evaluated the reaction characteristics of ammonia reduction in the gasified fuels, of non-catalytic denitration in exhaust, of air-fired gasified fueled combustions, and of H_2/O_2 stoichiometric combustion through experiments and full kinetic analyses [23], [25]-[27]. Results showed that the united scheme could describe the reaction characteristics in gasified fueled combustion and exhaust. On the other hand, various reaction schemes have been proposed worldwide for each reaction system including higher hydrocarbons. There was example of the GRI Mech 3.0 chemical kinetic mechanism used for calculation of the oxy-fuel gas turbine combustion [28]. But it need not be used since the gasified fuel contains a small percent of CH_4 and no C2 hydrocarbon.

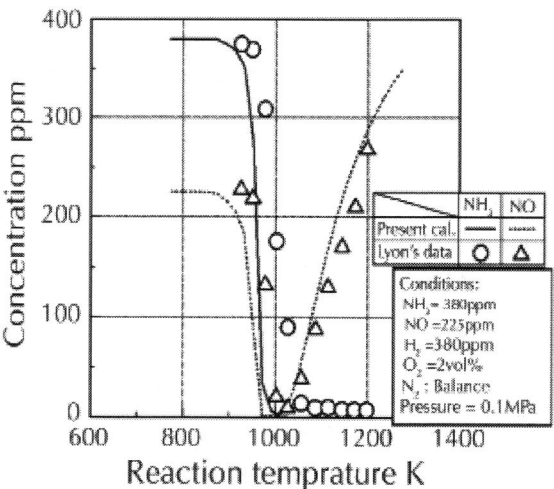

Figure 8: Comparison of kinetic analyses with experimental data of Lyon [24] on concentration of NH_3 and NO in the $NH_3 NOO_2 H_2$ system under conditions of selective no catalytic reduction of NOx.

We took thermodynamic data from the JANAF thermodynamics tables [29], and calculated the values of other species not listed in the tables based on the relationship between the Gibbs' standard energy of formation, G°, and the chemical equilibrium constant, K, obtaining a value of G° from the CHEMKIN database [30].

$$\Delta G° = R \times T \times \ln(K)$$

(2)

We in this study used the GEAR method [31] for numerical analysis as an implicit, multi-stage solution.

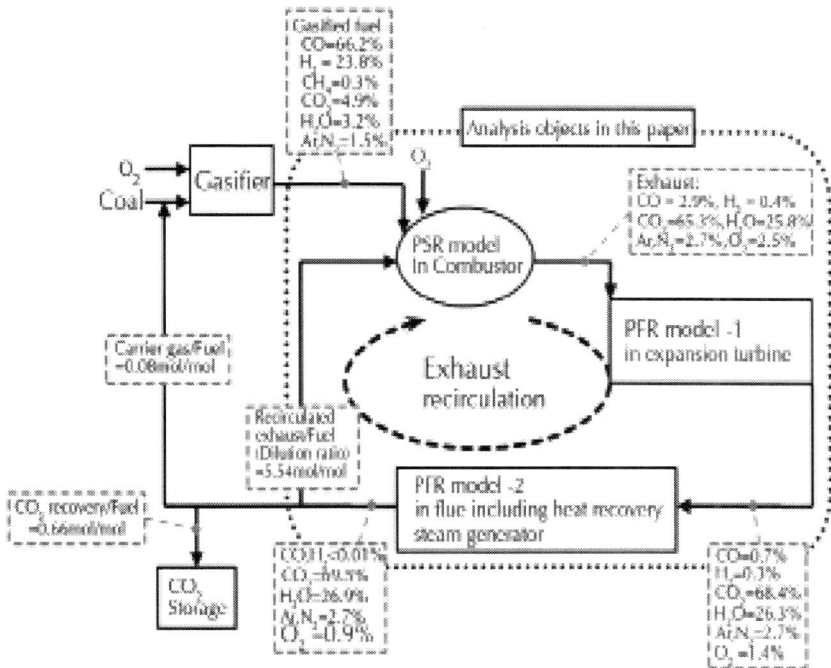

Figure 9: Schematic of algorism in semiclosed gas turbine for oxy-fuel IGCC under typical rated conditions. Recirculated gas turbine exhaust is injected into the PSR of the combustor alongside incoming stream of gasified fuel and oxidizer of O_2.

Furthermore, our algorithm is schematized in Figure 9. The model we employed in this study assumed all mixing processes to be ideal such that they could be represented by a combination of a perfectly-stirred reactor (PSR) and a plug flow reactor (PFR). When investigating the basic combustion reaction characteristics that were independent of combustor geometries, the combustor was simply modeled as the PSR. This combustor model was the simplest case of modular models employed by Pratt, et al. [32]. In the case of investigating the exhaust gas reaction characteristics in expansion turbine and flue, we employed the PFR model. Then, we employed a combination PSR and PFR model in order to explore the influence of exhaust recirculation on combustor emission characteristics and exhaust reaction characteristics in the semiclosed gas turbine.

CHARACTERISTICS OF STOICHIOMETRIC COMBUSTION WITH RECIRCULATING EXHAUST

Comparison with Air-fired Combustion

Figure 10 shows concentrations of principal chemical species against reaction time under the rated load conditions shown in Table 1, through a numerical analysis based on reaction kinetics with a PSR model of homogeneous reaction. Figure 11 also shows the principal chemical species against reaction time when burning CH_4 in the main components of natural gas with air under conditions where the reaction temperature is set at a constant value of the rated exhaust temperature of 1573K, and where the equivalence ratio is 0.32.

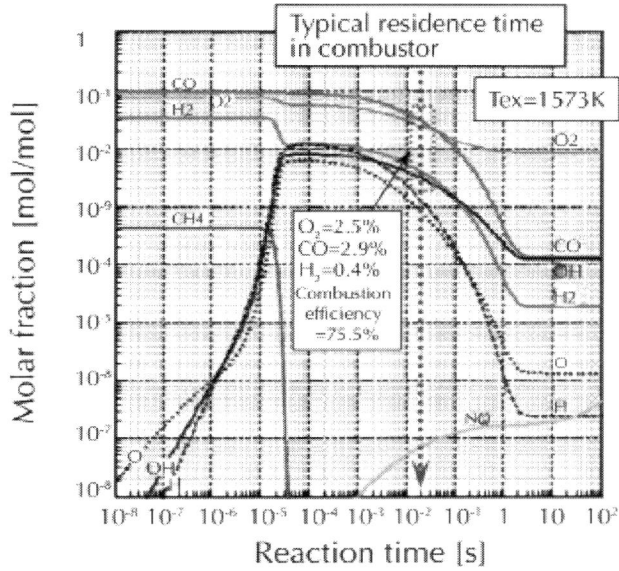

Figure 10: Chemical species behavior over time in gasified fuel/O$_2$ stoichiometric combustion with exhaust recirculation.

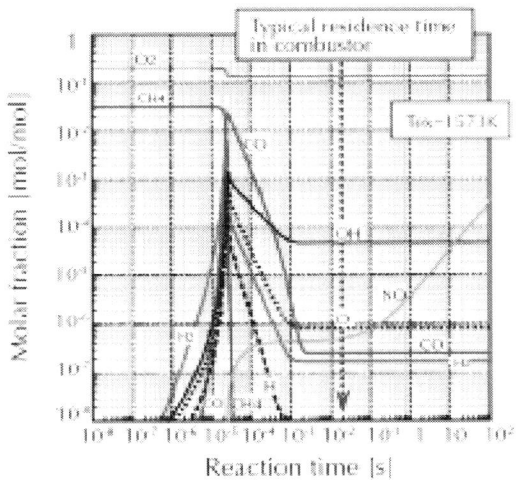

Figure 11: Chemical species behavior over time in conventional CH$_4$/air combustion.

In the case of burning gasified fuel under stoichiometric conditions with exhaust recirculation, fuel oxidation reaction proceeded slowly compared to that of conventional CH_4/air combustion. As a result, we found that CO and H_2 in exhaust remained unoxidized at around 2.9vol% and 0.4vol%, respectively, and residual O_2 at 2.5vol% in 20ms corresponded to the combustion gas residence time in the combustor. Combustion efficiency was estimated to remain at a low level of around 76%, compared with that of conventional industrial gas turbines.

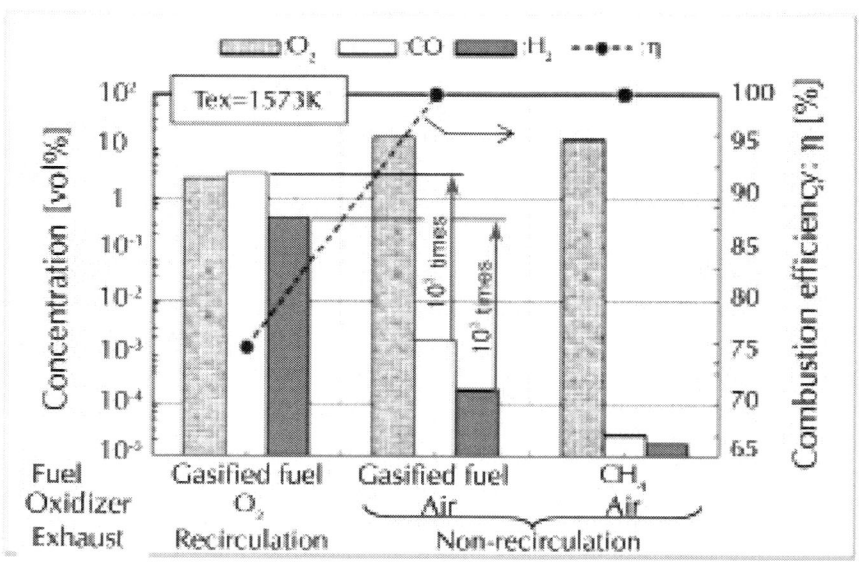

Figure 12: Comparison of emission characteristics with conventional air-fired combustions.

Figure 12 shows exhaust characteristics and combustion efficiencies at the combustor exit under the conditions of a 1573K combustor exhaust temperature, in comparison with the homogeneous premixed combustion of a gasified fuel/air and CH_4/air mixture. The stoichiometric combustion of gasified fuel/O_2 with exhaust recirculation causes a drastic decrease in combustion efficiency compared with the other two cases of air-fired combustion. We feel that it is therefore necessary to promote fuel oxidation, or to decrease combustible constituent CO emitted from the gas turbine.

Comparison to Each Oxy-fuel Gas Turbine Combustion

Figure 13 shows the results of numerical analysis in hydrogen/oxygen fired, stoichiometric combustion with exhaust recirculation of steam under the rated temperature conditions of 1573K. An overall equivalence ratio was set at 1 with other conditions equivalent to the rated conditions in Table 1.

Hydrogen/oxygen reaction began as rapidly as in the cases of gasified fuel/O_2 or CH_4/air fired combustion, shown in Fig.10 or Fig.11, respectively. After that, hydrogen oxidation reaction progressed faster than in the case of gasified fuel/O_2 fired combustion.

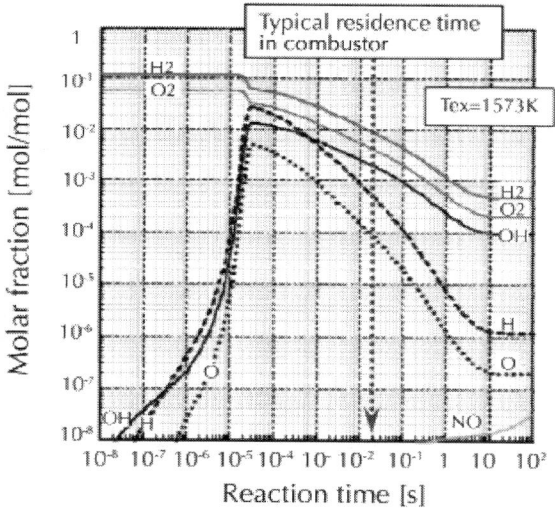

Figure 13: Chemical species behavior over time in H_2/O_2 combustion with steam recirculation.

Figure 14 shows exhaust characteristics and combustion efficiencies under the conditions of a 1573K combustor exhaust temperature compared with the cases of homogeneous premixed combustions of H_2/O_2 and CH_4/O_2 mixture with exhaust recirculation. Here, we set the composition of each recirculating exhaust to that of corresponding gas formed under equilibrium conditions.

As an example of oxygen-fired gas turbine using stoichiometric combustion with exhaust recirculation, Fig.14 also shows comparative calculations with test data of 1973K-class H_2/O_2 stoichiometric combustion with steam recirculation, conducted in the Japanese WE-NET project. Tests confirmed that the analytical results were almost in accordance with experimental results [33] concerning concentrations of residual O_2 constituent and unburned H_2 constituent in exhaust, and that the numerical analysis used in this study could estimate emission characteristics under conditions of achieved high combustion efficiency.

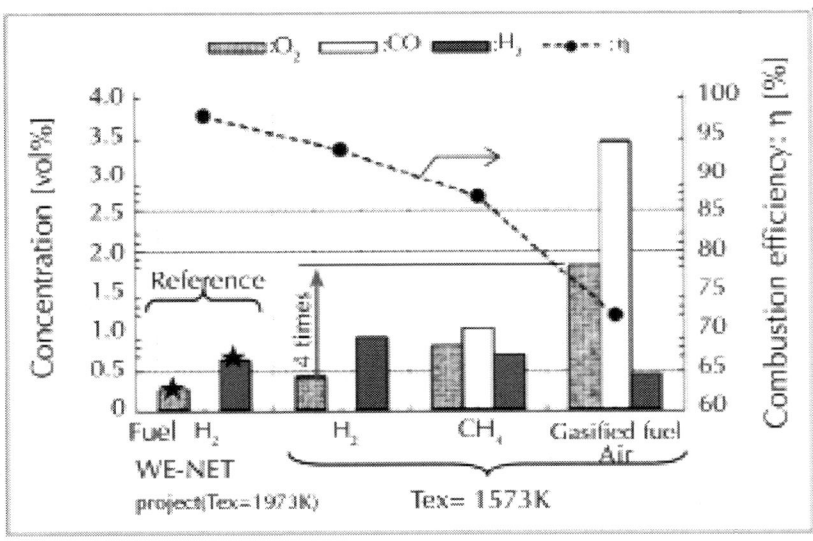

Figure 14: toichiometric combustion characteristics of each fuel; overall equivalence ratio is 1, markers (⋆)are test data under conditions where combustion pressure is set at 2.5MPa and recirculated steam temperature is 623K [33].

In the case of H_2/O_2 fired combustion such as in a hydrogen fired closed-cycle gas turbine, emissions of combustible constituent H_2 and residual O_2 in exhaust decreased 1vol% or below in a reaction time of 20ms, or combustion efficiency was estimated to reach up to around 93% at a temperature of 1573K. In the case of a CH_4/O_2 fired, closed cycle gas turbine, combustible constituent CO and residual O_2

concentration of combustor exhaust also decreased 1vol% or below, and combustion efficiency reached 87%, while combustible contents and residual O_2 emissions displayed a tendency to increase compared to the H_2/O_2 fired combustion.

Combustible contents and residual O_2 emissions, on the other hand, increased by several times in the case of gasified fuels compared with both cases of H_2 fired and CH_4 fired combustion. Combustion efficiency fell to a low level of 72%. In CO-rich fuel/O_2-fired combustion with exhaust recirculation, CO oxidation was strongly restrained by recirculating exhaust consisting mostly of CO_2 compared to other fuel constituents, and combustion efficiency was decreased. Therefore, to achieve highly efficient oxy-fuel IGCC, it is necessary to develop combustion control technologies of gasified fuel/O_2combustion with higher combustibility compared with the H_2/O_2 combustion technology in the WE-NET project or pre-combustion technologies.

Effects of Fuel Co/H$_2$ Molar Ratio on Emission Characteristics

Each quantity of CO and H_2 constituent in the gasified fuels differs chiefly according to the gasification methods, raw materials of feedstock, and water-gas-shift reaction as an optional extra for pre-combustion carbon capture system. Figure 15 shows influences of CO/H_2 molar ratio in the gasified fuel on the combustion emission characteristics with exhaust recirculation under the rated temperature condition of 1573K. In the case of varying the fuel CO/H_2 molar ratio under the conditions where the total amount of CO and H_2 constituent was set constant, dilution ratio (dilution gas/fuel molar ratio) was adjusted to maintain the combustion temperature at 1573K. Just like the case of Fig.14, overall equivalence ratio was set at 1, with other conditions equivalent to the rated conditions in Table 1. In the case of changing the fuel CO/H_2 molar ratio from 2.8 of base condition to 0.36, the amounts of CO and H_2 constituent replaced each other under the condition where the total amount of CO and H_2 was set constant of 90vol%.

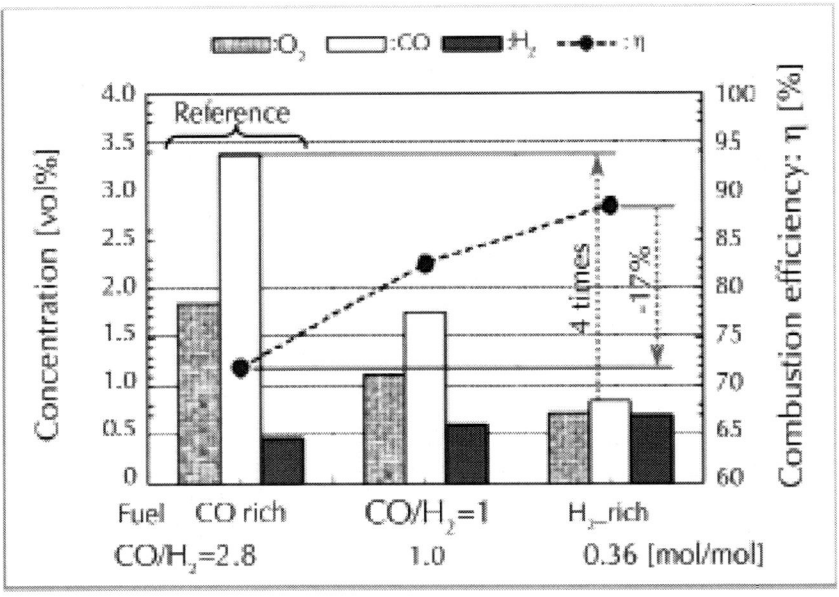

Figure 15: Effects of CO/H$_2$ molar ratio in fuel on stoichiometric combustion characteristics; overall equivalence ratio is 1. Notes: In the case of changing the fuel CO/H$_2$ molar ratio from 2.8 of base condition to 0.36, the amounts of CO and H$_2$ constituent replace each other under the condition where the total amount of CO and H$_2$ is set constant of 90vol%.

In the case of higher CO/H$_2$ molar ratio in the fuel, higher concentration of CO and lower concentration of H$_2$ in fuels increased CO emissions in combustion exhaust significantly, but have insignificant effects on reduction of H$_2$ emissions. As a result, in the case of CO rich gasified fuels, CO emissions increased four times those in the case of H$_2$ rich gasified fuel in the pre-combustion IGCC system, or combustion efficiency decrease by about 17%. This is explained both because H$_2$ is decomposed and produces OH, H and O radicals in the chain initiation as shown in Fig.10, and exhaust recirculation strongly inhibits oxidation of CO that is oxidized directly to CO$_2$ by the following reactions:

$$CO + OH \ (O+M, O_2, HO_2) \Leftrightarrow CO_2 + H \ (M, O, OH), \quad M:Thirdbody$$

$$(3)$$

Furthermore, H_2 is oxidized more rapidly than CO, or CO constituent controls overall oxidation reaction rate of fuel in the stoichiometric combustion with exhaust recirculation. Consequently, when the CO/H_2 molar ratio increased, CO oxidation rate and O_2 consumption rate decreased.

Effects of Equivalence Ratio on Emission Characteristics

Figure 16 shows the effects of an equivalence ratio on combustion emission characteristics under the rated temperature 1573K. When varying the equivalence ratio, the dilution ratio (dilution gas/fuel molar ratio) was adjusted to maintain the combustion temperature at 1573K. The horizontal axis indicated an apparent equivalence ratio, * calculated from fuel and an oxidizer without O_2 concentration in the dilution of recirculated exhaust. Emission features and combustibility of the combustor were characterized by combustion conditions near the burner. That is, Fig.16 indicated the influence of a so-called "local equivalence ratio" near the burner on combustion emission characteristics by using the apparent equivalence ratio *.

In the case of decreasing * from 0.98 to 0.95, combustion efficiency increased by only 5 percent, while overall equivalence ratio decreased from 0.89 to a low level of 0.75. That is, lowering the equivalence ratio could not result in any remarkable combustion promotion in CO-rich fuel/O_2 fired combustion with exhaust recirculation, while O_2 concentration in the exhaust significantly increased and the usage of exhaust to feed coal into the gasifier was restrained. It is necessary to decrease O_2 concentration in the carrier gas to feed coal by oxidation reactions using fuels such as hydrocarbons, or auxiliary power increased. Therefore, we have to decide the equivalence ratio in the combustor in consideration of the influence of residual O_2 on thermal efficiency of the whole system and performance of its equipments.

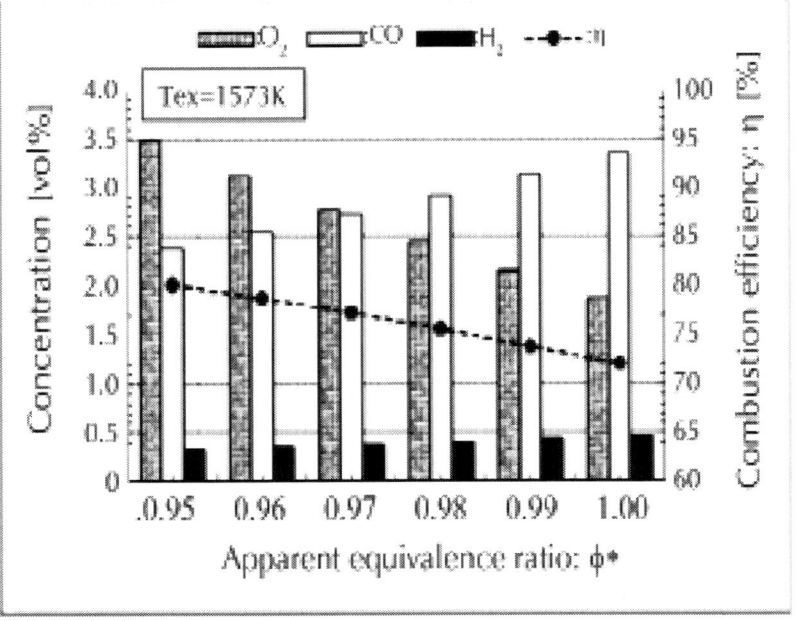

Figure 16: Effects of apparent equivalence ratio (*) on combustion emission characteristics.

Influences of Oxygen Concentration in Oxidizer on Emission Characteristics

O_2 concentration in oxidizer derived from an air separation unit differs according to the air separation and purification system. Figure 17 shows influences of oxygen concentration in oxidizer on the combustion emission characteristics under the rated temperature conditions. In the case of varying the oxygen concentration in oxidizer, dilution ratio (dilution gas/fuel molar ratio) was adjusted to maintain the combustion temperature at 1573K and overall equivalence ratio at 1.0. The remainder of the oxidizer without O_2 was set to N_2.

Emissions of residual O_2 and combustible constituents of CO and H_2 in exhaust tended to increase with the increase in oxygen concentration in oxidizer, or combustion efficiencies decreased. In the case of increasing O_2 concentration from 80vol% to 100vol%, combustion efficiency decreased by 13%, while residual concentrations of argon and nitrogen originated in air decreased. It was said that the non-condensable gases such as remaining O_2, argon and nitrogen resulted in increased condensation duty for the recovery of the CO_2 [21], or influence of residual constituents on the whole system and its equipments must be examined separately.

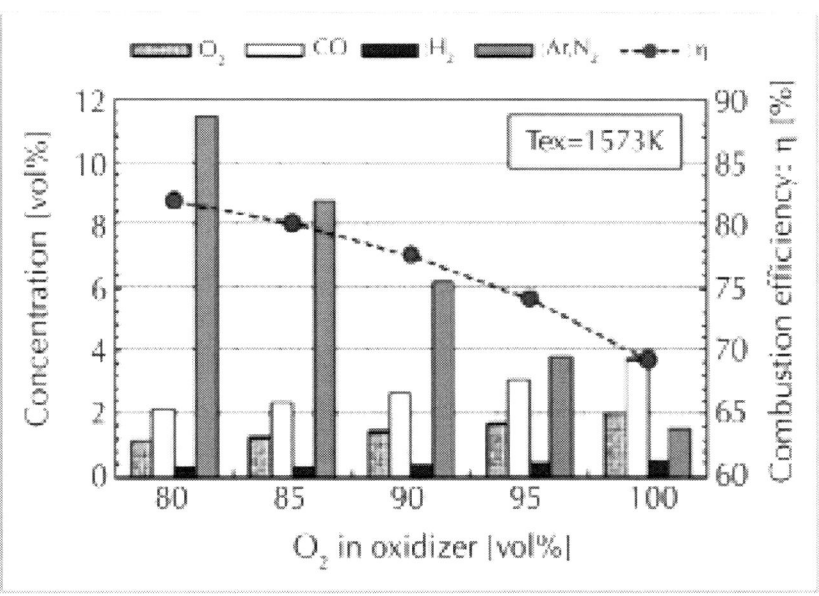

Figure 17: Effects of oxygen concentration in oxidizer derived from air separation unit on combustion emission characteristics; overall equivalence ratio is 1.

Reaction Characteristics of Gas Turbine Exhaust

Figure 18 shows a typical stream history of exhaust temperature and pressure from a gas turbine inlet to a compressor inlet of recirculating exhaust. Power was recovered from exhaust emanating from the combustor in the expansion turbine, and combustor exhaust temperature of 1573K with a pressure of 2.2MPa decreased to around

950K and 0.1MPa respectively at the turbine exit. Then, heat from expansion turbine exhaust was recovered through heat recovery steam generator (HRSG) in a flue, and exhaust temperature decreased to around 373K at the compressor inlet. In these analyses, we employed the PFR model for the turbine exhaust to the compressor inlet, and assuming that species in exhaust is evenly mixed and that there is no distribution of temperature and pressure in the mixtures. There is also no supply of added oxidizer and recirculating exhaust in the reaction processes.

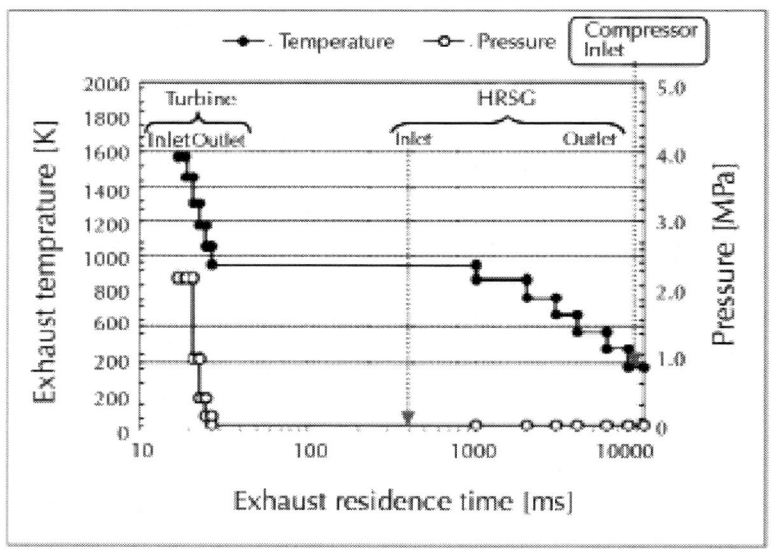

Figure 18: Typical stream history of exhaust temperature and pressure from gas turbine inlet to compressor inlet of recirculating exhaust.

Figure 19 shows the reaction characteristics of combustion gas in the combustor and exhaust gas from combustor outlet to compressor inlet of recycled exhaust using a combination PSR and PFR model. CO and H_2 at high concentration in exhaust could be slowed to oxidize under the temperature conditions of an expansion turbine and HRSG. Combustible constituents of CO and H_2, and residual O_2 therefore decreased in concentration along the exhaust flow direction. Oxidation reactions of CO and H_2 then nearly halted when the exhaust temperature decreased to 673K or less.

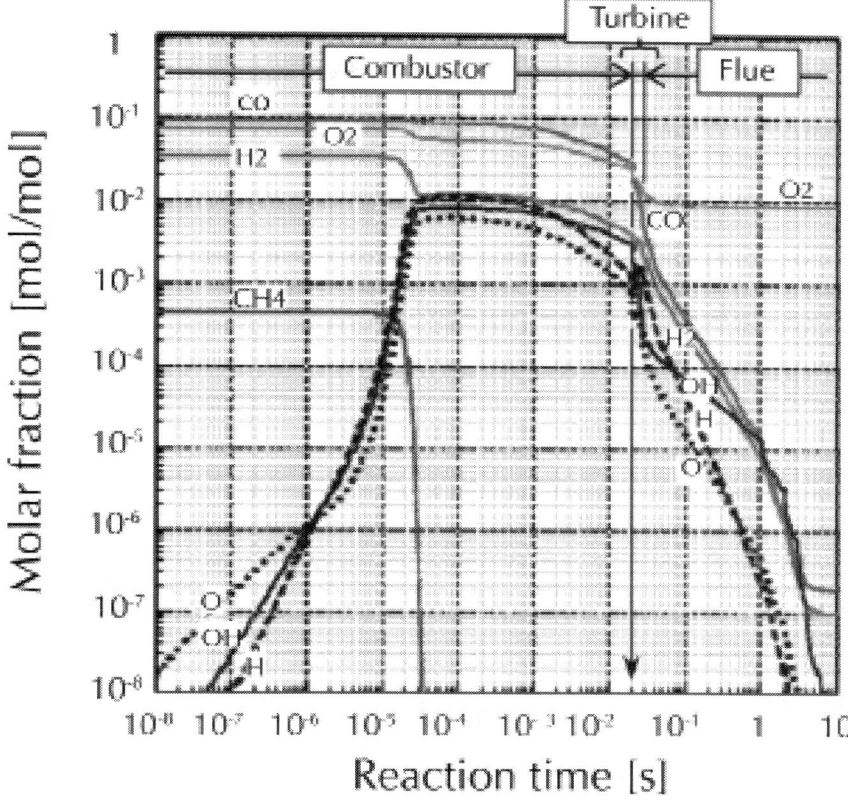

Figure 19: Chemical species behavior of combustion and exhaust gas over time in semiclosed cycle gas turbine, using PSR + PFR combined model. Combustor inlet conditions are the same as those in Fig.10 and flue includes HRSG.

Figure 20 shows emission characteristics of exhaust gases and combustion efficiencies at typical conditions based on the above reaction characteristics.

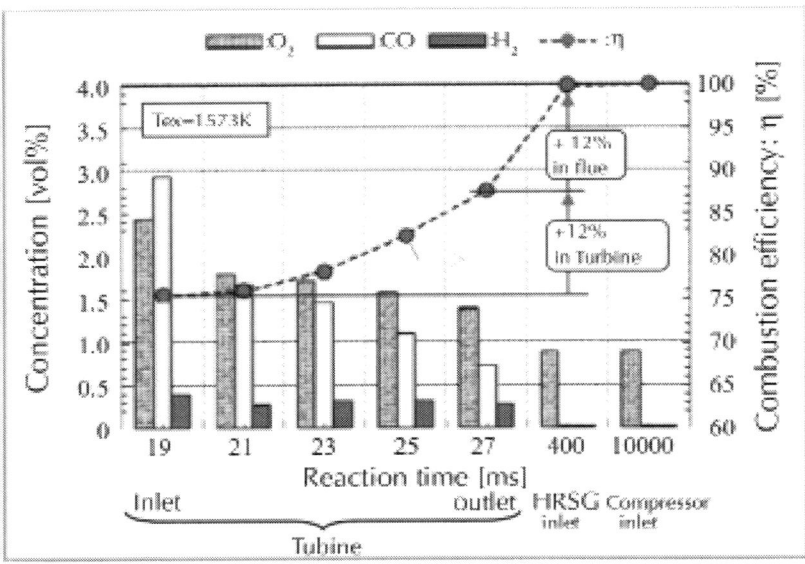

Figure 20: History of combustion emissions from gas turbine inlet to compressor inlet of recirculating exhaust

In a reaction time of 400ms corresponding to the residence time between combustor exit and HRSG inlet, combustible constituent CO and H_2 decreased less than 0.01vol%, and residual O_2 decreased to around 0.9vol%, while CO and H_2 concentration in combustor exhaust hovered at around 3vol% and 0.4vol%, respectively, and residual O_2 was at 2.3vol% under 20ms of typical combustion gas residence time in the combustor. Each oxidation reaction of combustible constituents in the turbine and the flue resulted in an increase in combustion efficiency of by about 12%, respectively, or a combustion efficiency was estimated to reach a high level of around 99.8%. If the reaction time was from 4 to 10 seconds when the exhaust gas temperature decreased to around 673K in the HRSG, both combustible constituent CO and H_2 decreased to 10ppmv, or the combustion efficiency reached 100%.

From the abovementioned results, we were able to clearly slow combustible constituents in expansion turbine exhaust to oxidize under conditions of exhaust temperatures of over 673K, or recover burning energy from unburned fuel in HRSG. However, in this case, since reaction heat of combustible constituents in HRSG corresponds to a fuel for reheat type HRSG, some of the supplied fuel could not devote enough energy to a combined cycle thermal efficiency. In order to achieve highly efficient oxy-fuel IGCC, it was therefore necessary to increase combustion efficiency as much as possible in the gas turbine combustor.

Influences of Exhaust Recirculation on Thermal-NOX Emissions

As shown in Fig.20, it is found that combustible constituents reached almost equilibrium concentration at compressor inlet of recirculated exhaust, or that equilibrium gases were working fluids in the semi closed cycle gas turbine. On the other hand, NOx constituents increased by exhaust recirculation and were saturated by a balance between exhaust recirculation and CO_2 recovery process. Figure 21 demonstrates the influence of exhaust recirculation on thermal-NO emission characteristics through numerical analyses of a combination PSR and PFR model as in the case of Fig.20, with compositions of fuel and oxidizer shown in Table 1. In these analyses, Fig.21 indicates the direct effects of recirculating NO constituent on NO-saturating concentration under conditions where dilution gas composition without NO constituent is constant. That is, we repeated the calculation of one exhaust-recirculating loop shown in Fig.20, and investigated influence of exhaust recirculation on oxidation-reduction reaction of NO through full kinetic analyses. Combustion temperature was set at 1623K, and pressure at 3.0MPa; a little higher than indicated in Table 1. Dilution gas of recirculated exhaust at combustor inlet were set to equilibrium composition.

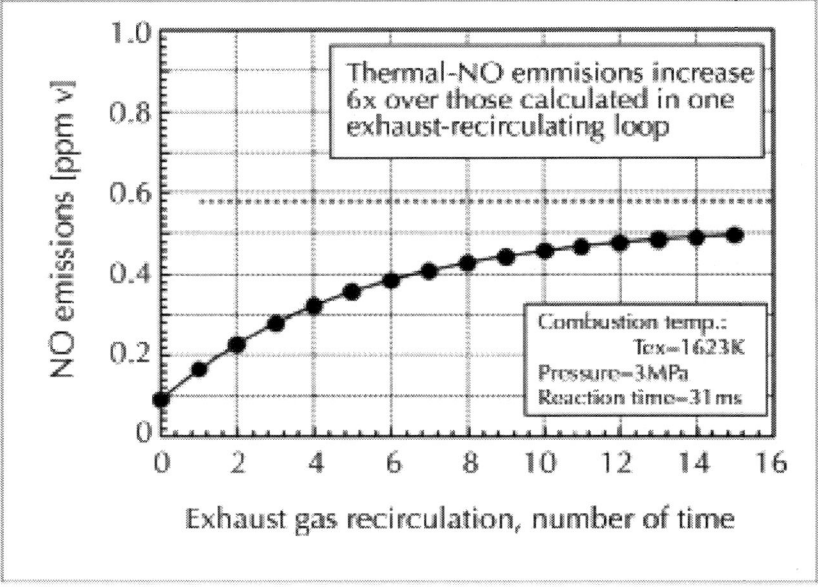

Figure 21: Influence of exhaust recirculation on thermal-NO emissions.

Thermal-NO emissions increased in response to a number of times exhaust was recirculated and reached around 6 times higher than those calculated in one exhaust-recirculating loop, while NO production itself was not significantly large due to the small component amounts of N_2, as shown in Table 1. However, since thermal-NO production depends on reaction temperature, or thermal-NO emissions are strongly affected by both mixing processes and non-uniformities of mixtures, we need further studies on thermal-NO emissions in the processes of combustion and combustor design.

CONCLUSIONS

Oxy-fuel IGCC employing an oxygen-fired, semi closed cycle gas turbine with exhaust recirculation enables the realization of highly-efficient, zero-emissions power generation. Numerical analyses in this paper showed both combustion emission characteristics of the semi closed cycle gas turbine combustor and oxidations of unburned fuel constituents in the turbine exhaust in a flue, compared with conventional air-fired gas turbines and advanced O_2-fired gas turbines. As a result, we were able in this study to clarify that unburned constituents in combustor exhaust were slow to oxidize under temperatures of over 673K in the flue and that all fuel energy could be used for power generation, while the oxidation reaction of CO-rich gasified fuel under stoichiometric conditions could be restrained with CO_2 constituents in re-circulated exhaust at decreased combustion efficiency. In this case, however, all the supplied fuel could not devote enough energy to boosting combined cycle thermal efficiency, leading therefore to a decrease in thermal efficiency overall. As a next step, we propose the need to promote oxidation reaction by developing combustion control technology for the improvement of plant thermal efficiency.

ACKNOWLEDGEMENTS

The author wishes to express their appreciation to the many people who have contributed to this investigation.

REFERENCES

1.   BP Statistical Review, "Historical Statistical Data from 1965-2010" and "BP Statistical Review of World Energy 2011", http://www.bp.com/statisticalreview/ (accessed on 13 March 2011).

2.   for example, The Energy Data and Modelling Center, The Energy Conservation Center, Japan, 2010, "2010 EDMC Handbook of Energy & Economic Statistics in Japan", ISBN:978-4-87973-365-8.

3.   Shirai,H., et al., 2007, "Proposal of high efficient system with CO2 capture and the task on integrated coal gasification combined cycle power generation," Central Research Institute of Electric Power Industry (CRIEPI) Report No.M07003. (in Japanese)

4.   Nakao,Y., et al., 2009, "Development of CO2 capture IGCC system -Investigation of aiming at higher efficiency in CO2 capture IGCC system-," CRIEPI report No.M08006. (in Japanese)

5.   Hasegawa,T., et al., 2011, "Study on gas turbine combustion for highly-efficient IGCC power generation with CO2 capture -2nd report: emission analysis of gasified-fueled gas turbines with

circulating exhaust & stoichiometric combustion-," CRIEPI report No.M10005, ISBN:978-4-7983-0462-5. (in Japanese)

6. Hasegawa,T., 2012, "Combustion Performance in a Semi-Closed Cycle Gas Turbine for IGCC Fired with CO-Rich Syngas and Oxy-Recirculated Exhaust Streams," Trans. ASME, J. Eng. Gas Turbines Power, Vol.134?, Issue *, pp.***-***, ISSN:0742-4795. (in press)

7. Kobayashi,M., et al., 2009, "Optimization of dry desulfurization process for IGCC power generation capable of carbon dioxide capture -determination of carbon deposition boundary and examination of countermeasure-," CRIEPI report No.M09015. (in Japanese)

8. Umemoto, S., et al., 2010, "Modeling of coal char gasification in coexistence of CO2 and H2O," Proceedings of The 27th Annual International Pittsburgh Coal Conference, The University of Pittsburgh, Hilton Istanbul, Istanbul, TURKEY (13 October 2010).

9. Kidoguchi, K., et al., 2011, "Development of oxy-fuel IGCC system with CO2 recirculation for CO2 capture -experimental examination on effect of gasification reaction promotion by CO2 enriched using bench scale gasifier facility-," Proceedings of the ASME Power Conference 2011 and the International Conference on Power Engineering 2011 (14 July 2011).

10. Engineering Advancement Association of Japan, WE-NET Home Page/WE-NET report, http://www.enaa.or.jp/WE-NET/report/report_j.html (accessed on 12 March 2011).

11. Ciferno,J.P., et al., 2010, "DOE/NETL carbon dioxide capture and storage RD&D roadmap (December 2010)," available online: http://www.netl.doe.gov/technologies/   carbon_seq/refshelf/CCSRoadmap.pdf (accessed on 13 September 2011).

12. Ciferno,J.P., et al., 2011, "DOE/NETL Advanced CO2 Capture R&D Program: Technology Update, May 2011 Edition," available online: http://www.netl.doe.gov/   technologies/coalpower/ewr/pubs/CO2CaptureTechUpdate051711.pdf (accessed on 1 November 2011).

13. Bancalari,Ed., Chan,P., and Diakunchak,I.S., 2007, "Advanced hydrogen gas turbine development program," Proceedings of the ASME Turbo Expo 2007: Power for Land, Sea and Air GT2007, ASME paper GT2007-27869, Montreal, Canada, May 14-17, 2007.

14. Todd,D.M. and Battista,R.A., 2000, "Demonstrated applicability of hydrogen fuel for gas turbines," Proceedings of the IChemE "Gasification 4 the Future" Conference, Noordwijk, the Netherlands, April 11-13, 2000.

15. Kajitani,S., Suzuki,N., Ashizawa,M., and Hara,S., 2006, "CO2 gasification rate analysis of coal char in entrained flow coal gasifier," Fuel 85, pp.163-169.

16. Nunokawa,M., Kobayashi,M., Shirai,H., 2008, "Halide compound removal from hot coal-derived gas with reusable sodium-based sorbent," Powder Technology, 180(1-2), pp.216-221.

17. Akiho,H., Kobayashi,M., Nunokawa,M., Tochihara,Y., Yamaguchi,T., Ito,S., 2008, "Development of Dry Gas Cleaning System for Multiple Impurities -Proposal of low-cost mercury removal process using the reusable absorbent - ," Central Research Institute of Electric Power Industry, Report No.M07017. (in Japanese)

18. Ozawa,Y. and Tochihara,Y., 2010, "Study of Ammonia Decomposition in Coal Derived Gas - Decomposition Characteristics over Supported Ni Catalyst -," Central Research Institute of Electric Power Industry, Report No.M09002. (in Japanese)

19. Nunokawa,M., Kobayashi,M., Nakao,Y., Akiho,H., Ito,S., 2010, "Development of gas cleaning system for highly-efficient IGCC -Proposal for scale-up scheme of optimum gas cleaning system based on generating efficiency analysis-," Central Research Institute of Electric Power Industry, Report No.M09016. (in Japanese)

20. for example; Hasegawa,T., 2006, "Gas turbine combustor development for gasified fuels and environmental high-efficiency utilization of unused resources," Journal of the Japan Petroleum Institute, Vol.49, No.6, pp.281-293.

21. Li,H., Yan,J., and Anheden,M., 2009, "Impurity inpacts on the purification process on Oxy-fuel comubtion based on CO2 capture and storage system," Appl. Energy2009; 86, pp.202-213.

22. Miller,J.A. and Bowman,C.T., 1989, "Mechanism and modeling of nitrogen chemistry in combustion," Prog. Energy Combust. Sci., vol.15, pp.287-338.

23. Hasegawa,T., et al., 1998, "Study of ammonia removal from coalgasified Fuel," Combust. Flame 114, pp.246258.

24. Lyon,R.K., 1979, "Thermal DeNOx: how it works," Hydrocarbon Processing 1979 (ISSN 0018-8190), Gulf Publishing Co., Houston, Texas, (October 1979), vol.59, No.10, 109-112.

25. Hasegawa,T., Sato,M., Sakuno,S., and Ueda,H., 2001, "Numerical Analysis on Apprication of Selective Non Catalytic Reduction to Wakamatsu PFBC Demonstration Plant", Paper number : JPGC2001/FACT-19052, Presented at the 2001 International Joint Power Generation Conference, New Orleans, Louisiana, U.S.A., 2001-6-6.

26. Nakata T., Sato M., and Hasegawa T., 1998, "Reaction Kinetics of Fuel NOx Formation for Gas Turbine conditions", Trans. ASME, J. Eng. Gas Turbines Power, Vol.120, No.3, pp.474-480.

27. Hasegawa,T., et al., 1997, "Fundamental Study On Combustion Characteristics Of Gas Turbine Using Oxygen-Hydrogen -Numerical Analysis Of Formation/Destruction Characteristics Of Hydrogen, Oxygen And Radicals-," CRIEPI report No.W96007. (in Japanese)

28. C.Y. Liu, G. Chen, N. Sipöcz, M. Assadi, X.S. Bai, 2012, "Characteristics of oxy-fuel combustion in gas turbines," Applied Energy, Volume 89, Issue 1, January 2012, pp.387-394.

29. Chase, Jr.M.W.; Davies, C.A.; Downey, Jr.J.R.; Frurip, D.J.; McDonald, R.A.; Syverud, A.N., 1985, "JANAF Thermodynamical Tables, 3rd Edition.," J. Phys. Chem. Reference Data, Vol.14.

30. Kee,R.J., Rupley,F.M., and Miller,J.A., 1990, "The CHEMKIN Thermodynamic Data Base," Sandia Report, SAND 878215B.

31. Hindmarsh,A.C., 1974, "GEAR: Ordinary differential equation system solver," Lawrence Livermore Laboratory, Univ. California, Report No. UCID30001, Rev.3.

32. Pratt, D.T.; Bowman, B.R.; Crowe, C.T. Prediction of Nitric Oxide Formation in Turbojet Engines by PSR Analysis. AIAA paper 1971, No.71–713.

33. Engineering Advancement Association of Japan, WE-NET Home Page/WE-NET report, http://www.enaa.or.jp/WE-NET/report/1998/japanese/gif/823.htm#823 (accessed on 12 March 2012).

Chapter 7

Gas Turbine Cogeneration Groups Flexibility to Classical and Alternative Gaseous Fuels Combustion

Ene Barbu[1], Romulus Petcu[1], Valeriu Vilag[1], Valentin Silivestru[1], Tudor Prisecaru[2], Jeni Popescu[1], Cleopatra Cuciumita[1], and Sorin Tomescu[1]

[1]National Research and Development Institute for Gas Turbines COMOTI, Bucharest, Romania

[2]Politehnica University, Bucharest, Romania

INTRODUCTION

The gas turbine installations represent one of the most dynamic fields related to the applicability area and total installed power. The gas turbines have been developed particularly as aviation engines but they find their applicability in many areas, one of which being simultaneously obtaining electric and thermal energy in gas turbine cogeneration plants. The gas turbine cogeneration plants may be classified based on the constructive technology of the gas turbine in

[1]: aeroderivative gas turbines plants (up to 10 MW); industrial gas turbines plants, specifically designed for obtaining energy (from 10 up to hundreds MW). An aviation gas turbine with expired flying resource is still functional due to the fact that the flight time is limited as a consequence of the specific safety normatives requirements. Therefore the aeroderivative gas turbine is defined as a gas turbine, derived from an aviation gas turbine, dedicated to ground applications. According to the initial destination, these gas turbines have been designed for maximum efficiency considering the limited fuel quantity available for an aircraft flying large distances. The basic idea in developing the aeroderivative gas turbine has been to transfer all the scientific and technologic knowledge ensuring a high degree of energy utilization (design concepts, materials, technologies, etc.) from aviation to ground [2]. Therefore the obtained gas turbines are lighter, with smaller size, increased reliability, reduced maintenance costs and high efficiency. The remaining resource for ground applications is proportional with the flight resource, being able to reach up to 30,000 hours considering the lower operating regimes. From the point of view of the actual application, the free power turbine groups are the most recommended [3]. Unlike the aeroderivative turbine power units, the industrial power units are built by the original producer with the necessary changes for actual industrial application. The development of aeroderivative and industrial gas turbines has been affected by the progress of the aviation gas turbines in military and civilian fields. Many aeroderivative gas turbines ensure compression rates of 30:1 [4]. The industrial gas turbines are cumbersome but they are more adaptable for long running and allow longer periods between maintenance controls. The base fuel for gas turbine cogeneration groups is the natural gas (with a possible liquid fuel as alternative) but the diversification of the gas turbines users and the increase in fuels price has pushed the large producers to consider alternative solutions. Nowadays the most utilized fuels in gaseous turbines are the liquid and gas ones (classic and alternative). The high temperature of the exhausted gases, approximately 590 °C on some gas turbines, allows the valorization of the heat resulted in a heat recovery steam generator. Due to the fact that the oxygen concentration in the exhausted gases is 11-16% (volume), a supplementary fuel burning may be applied (afterburning) in order to increase the steam flow rate, compared to the case of the heat recovery steam generator [5]. The afterburning leads to an increase in flexibility and global efficiency of

the cogeneration group, allowing the possibility to burn a large variety of fuels, both classic and alternative. Nitrogen oxides usually represent the maine source of emissions from gas turbines. The NO_x emissions produced by the afterburning installation of the cogeneration group are different according to the system, but they are usually small and in some cases the installation even contributes to their reduction [6]. The usual methods for NO_x emissions reduction, water or steam injection for flame temperature decrease, affect the gas turbine performances, particularly to high operating regimes, leading to CO emissions increase. It must be noted that the load of the gas turbine also affects the emissions, the gas turbine being designed to operate at high loads. The general theme of the chapter is given by the technological aspects that must be considered when aiming to design a gas turbine cogeneration plant flexible from the points of view of the utilized fuel and the qualitative and quantitative results concerning some classic and alternative gas fuels. Based on the specific literature in the field and the experience of National Research and Development Institute for Gas Turbines COMOTI Bucharest, there are approached theoretic and experimental researches concerning the utilization of natural gas, as classic fuel, and respectively dimethylether (DME), biogas (landfill gas) and syngas, as alternative fuels, in gas turbine cogeneration groups, the interference between flexibility and emissions. It is particularly analysed the issue of reutilization of aviation gas turbines in industrial purposes by their conversion from liquid fuel to gas fuels operation. There is further presented the actual method of conversion for an aviation gas turbine in order to be used in cogeneration groups.

THE AERODERIVATIVE GAS TURBINE – A SOLUTION FOR GAS TURBINE COGENERATIVE GROUPS FLEXIBILITY ON GAS FUELS

The flexibility of the gas turbine cogeneration plants implies reaching an important number of requirements: operating on classic and alternative fuels; capability of fast start; capability to pass easily from full load to partial loads and back; maintaining the efficiency at full load and partial loads; maintaining the emission to a low level even when operating on partial loads. Internationally, many companies with top performance in aviation gas turbines are involved in aeroderivative programs in response to market demands for energy producing installations. The best known among these are: Rolls-Royce, Pratt & Whitney, General Electric, Motor Sich, Turbomeca, MTU, etc. Rolls-Royce has developed the RB 211-H63 gas turbine starting from the aviation RB 211 which, through novel constructive and technologic transformations has been pushed to efficiency up to 41.5%. A 38 MW version will be available in 2013 with the possibility of upgrade to 50 MW in future years [7]. Many gas turbine producers aim to reach the full load in ten minutes from the start. A Japanese project of Mitsubishi Heavy Industries Ltd. (MHI) aims to manufacture a gas turbine operating at 1700 °C inlet temperature and 62 % efficiency. Pratt & Whitney, starting from the PW 100 turboprop, have developed the ST aeroderivative gas turbine family (ST 18, ST 40). The researches conducted at National Research and Development Institute for Gas Turbines COMOTI Bucharest have allowed obtaining aeroderivative gas turbines in the 20 – 2,000 kW range, through valorisation of the aviation gas turbines with exhausted flight resource, obsolete or damaged. Therefore the AI 20 GM (figure 1, right) aeroderivative turboshaft, operating on natural gas, is based on the AI 20 turboprop (figure 1, left). The AI 20 GM is used in power groups driving the backup compressors in the natural gas pumping stations on the main line at SC TRANSGAZ SA. The aeroderivative GTC 1000 (figure 2, right), based on TURMO IV C (figure 2, left), operating on natural gas, is used in a power group driving two serial centrifugal compressors for the compression of the associated drill gas, in one SC OMV PETROM SA oil exploitation, at Ţicleni – Gorj. Researches have also been conducted regarding the valorisation of the landfill gas in a aeroderivative gas turbine applicable to cogeneration groups [2]. A project for a cogeneration application using the GTE 2000 aeroderivative gas turbine has been started in 2000. The result of the project is a cogeneration plant, with two independent lines, producing electric and thermal (hot water) energy, located in the municipality of Botosani, with SC TERMICA SA as beneficiary (figure 3, left). The experience acquired from the GTE 2000 cogeneration plant has been used in a new project for a medium power aeroderivative gas turbine cogeneration plant, the application using the ST 18 A aeroderivative gas turbine, manufactured by Pratt & Whitney. The ST 18 A aeroderivative gas turbine has been derived from the aviation PW 100 through redesigning a series of components of which are distinguished the combustion chamber, the case and the intake. Furthermore, the ST 18 A has been designed and manufactured to operate with water injection in the combustion chamber (duplex burners), method that ensures the reduction of NO_x emissions. The application consists in a cogeneration plant, with two independent cogeneration lines, producing electric and thermal (superheated steam used in the oil extraction technologic process) energy. The beneficiary of the application is SC OMV PETROM SA, Suplacu de Barcău, Bihor County (figure 3, right) [8]. What makes the difference between aviation and aeroderivative gas turbines are operating conditions and reliability. Thus, aviation gas turbines have over the period of their useful life so many ordered starts and stops (associated with aircraft flight), short operation between starting and stopping (of hours), short periods between revisions (after each stop) and overhauls (after more than 1,000 hours of operation), the lifespan of about 12,000 cumulative hours of operation. Aeroderivative gas turbines can operate up to 8,000 hours continuously without ordered stop, overhauls are made at intervals up to 30,000 cumulative operating hours and, for some brands, the cumulative operating ranges may be even higher.

Figure 1: AI 20 turboprop (left) and AI 20 GM gas turbine (right) [2].

 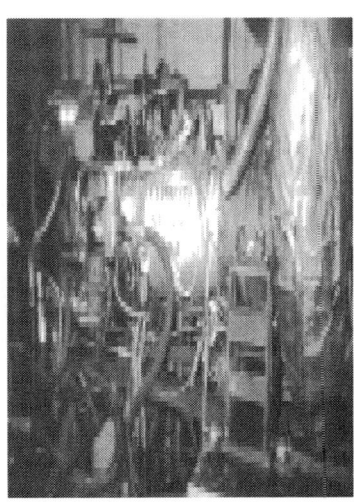

Figure 2: TURMO IV C turboshaft (left) and GTC 1000 gas turbine (right) [2].

Figure 3: GTE 2000 – Botoşani (left) and 2xST 18 – Suplacu de Barcau (right) plants.

Classic and Alternative Fuels for Gas Turbine Cogeneration Groups

The performances of the gas turbine cogeneration groups (efficiency and emissions) depend in high degree of the type and physical and chemical properties of the used fuels. Depending on the lower heating value (LHV), in relation to natural gas (LHV=30-45 MJ/Nm^3), typical gas fuels can be classified as [9]: high heating value (LHV=45-190 MJ/Nm^3; butane, propane, refinery off-gas), medium heating value (LHV=11.2-30 MJ/Nm^3; weak natural gas, landfill gas, coke oven gas), low heating value (LHV<11.2 MJ/Nm^3; BFG - Blast Furnace Gas, refinery gas, petrochemical gas, fuels resulted through gasification etc).

General Requirements Regarding the Utilization of Fuels in Gas Turbines

Figure 4: Control – measurement station for natural gas at 2xST 18 – Suplacu de Barcau plant (left) with booster (right) 1 – cogeneration power plant; 2 – control – measurement station; 3 – booster.

For the gas turbines used in cogeneration groups, for economic reasons, the most used fuels are heavy oil and waste products from various manufacturing or chemistry processes [3]. Using liquid fuels imposes: ensuring combustion without incandescent particles and deposits on the firing tube and the turbine; decreasing the corrosive action of the burned gases caused by the aggressive compounds (sulphur, lead, sodium, vanadium, etc.); solving the pumping and atomization issues (filtration, heating, etc.). A series of fuels must be well purified or filtrated for eliminating water, solid particles or some remiss substances. Heavy liquid fuels must be heated to a convenient temperature to allow their proper pumping and spraying. Coke number and tar number are of particular interest for burning in gas turbines. Coke number (carbon residue) represents the residue left by an oil product (fuel oil, diesel, etc.) when burned in special conditions (closed space, restricted air access, etc.), expressed in mass percent. Tar number indicates the presence of resins, aromatic hydrocarbons, etc. but it must be considered for information only. In order to define the combustion behaviour of a heavy liquid fuel (like oil) it would be indicated to consider as a criterion the product of the coke number and tar number [10]. In terms of reusing aviation gas turbines in industrial purposes, the possibilities of using liquid fuels are decreasing. For each application, the requirements of the beneficiary must be analysed related to the characteristics of the fuels affecting the combustion (density, molecular weight, evaporation limit, flammability temperature, volatility, viscosity, surface tension, latent heat of vaporization, calorific value, the tendency for soot, etc.). In terms of using gas fuels, the problem is less challenging due to their thermal stability, high heating value, lack of soot and tar. However, in order to ensure the pressure level required by the gas turbine, afterburning, etc., the elimination of water and different impurities, a control – measurement station must be provided for the gas fuels to be used (natural gas at 2xST 18 plant – figure 4). Some alternative gas fuels (resulted through gasification and biomass pyrolysis), biogas, residual gases from industrial processes (rich in hydrogen) can play an important role in the operation of the gas turbine cogeneration groups, but they must reach some requirements regarding the calorific value and the composition [11]. Therefore there is necessary to eliminate the impurities, tar, to limit the sulphur and its compounds to 1 mg/Nm^3, respectively the alkaline metal compounds to 0.1 mg/Nm^3 [12].

Alternative Fuels – Characteristics and Consequences Regarding Their Use in Gas Turbine Cogeneration Groups

The biogas produced through anaerobic fermentation is cheap and constitutes a renewable energy source producing, from burning, neutral carbon dioxide (CO_2) and offering the possibility of treatment and recycling for residues and secondary agricultural products, various biowaste, organic waste water from industry, sewage and sewage sludge. The properties and the composition of biogas are different depending on the raw material used, processing system, temperature, etc. The comparative compositions of natural gas and biogas are given in table 1 [13]. For both fuels the main component (giving the energetic value) is the methane (CH_4), the significant differences being given by the high content of CO_2 and H_2S (hydrogen sulphide) in biogas. Technically, the main difference is given by the Wobbe index for natural gas (see chapter 2.2), two times higher than the index for biogas. This leads to a limited possibility of replacing the natural gas with biogas because only gases with similar Wobbe index can substitute each other. The improvement of the biogas can be achieved by replacing CO_2 with CH_4 so as to approach the characteristics of natural gas. Furthermore, the water and hydrogen sulphide must be eliminated to avoid the harmful action of the resulted sulphuric acid on different components of the cogeneration group (gas turbine, afterburning installation, heat recovery steam generator, etc.). Landfill gas resulted from waste deposits represents a cheap energy source, with a composition similar to the biogas resulted from anaerobic fermentation (45-60 % methane, 40-55 % carbon dioxide) [2]. When it comes to using biogas in gas turbine cogeneration groups or introducing it in the natural gas network, special treatment is required (condensate separation, drying, adsorption of volatile substances, etc.). Dimethylether (DME, CH_3-O-CH_3) is a clean alternative fuel which can be produced from fossil fuels, namely coal or vegetal biomass gasification. It can be transported and stored similar to liquefied petroleum gas (LPG), its physical and chemical characteristics, related to natural gas in Ardeal (99.8 % CH_4 and 0.2 % CO_2), being given in table 2 [14]. The flame produced by burning the dimethylether is very similar to the flame produced by the natural gas (figure 5), which makes it suitable to be used as fuel in transportation, cogeneration groups, etc.

Through biomass of coal gasification (with oxidant agents such as oxygen, air, steam, etc.) it can obtain synthesis gas (syngas) with main components hydrogen (H_2) and carbon monoxide (CO). The syngas can be used to obtain methanol, hydrogen, methane, etc. or can be used as fuel in gas turbine cogeneration groups. Since leaving the gas-producing installation the gas containes ash particles and various compounds of chlorine, fluorine, alkali metals, etc., which must be removed to protect the cogeneration line. Through gasification of different biomass categories and utilization of different gasification technologies, the composition of the resulted gas and the lower heating value (LHV) can vary according to tables 3 and 4 [12, 15]. Tables 1 and 3 show that the lower heating values for biogas and syngas are lower than for the natural gas, requiring, in their application in cogeneration groups, higher mass flow rates with minimum pressure losses. Therefore, the injection nozzles of the gas turbine and the burners of the afterburning installation must be designed for velocities allowing a homogenous mixture between fuel and oxid, as well as low pressure losses. The syngas contains high quantities of hydrogen which affect the combustion in gas turbine cogeneration groups in terms of flame stability, combustion efficiency, etc. Using hydrogen as fuel and introducing a component with dilution role (steam, nitrogen, etc.) the operation of the gas turbine is affected [16].

Table 1: Composition, physical and chemical proprieties for natural gas and biogas [13]

No.	Name	Natural gas	Biogas
1	CH4 [vol %]	91.0	55-70
2	CnH2n [vol %]	8.09	0
3	CO2 [vol %]	0.61	30-45
4	N2 [vol %]	0.3	0 - 2
5	Lower heating value [MJ/Nm3]	39.2	23.3
6	Density [kg/Nm3]	0.809	1.16

Table 2: Physical and chemical characteristics for natural gas (Ardeal) and dimethylether [14]

No.	Name	Natural gas (Ardeal)	Dimethylether
1	Theoretical combustion temperature [0C]	1,900	2,000
2	Autoignition temperature [0C]	650-750	350
3	Lower heating value [MJ/Nm3]	35.772	59.230
4	Explosion limit [% gas in air]	5 - 15.4	3 - 18.6
5	Density [kg/Nm3]	0.716	2.052

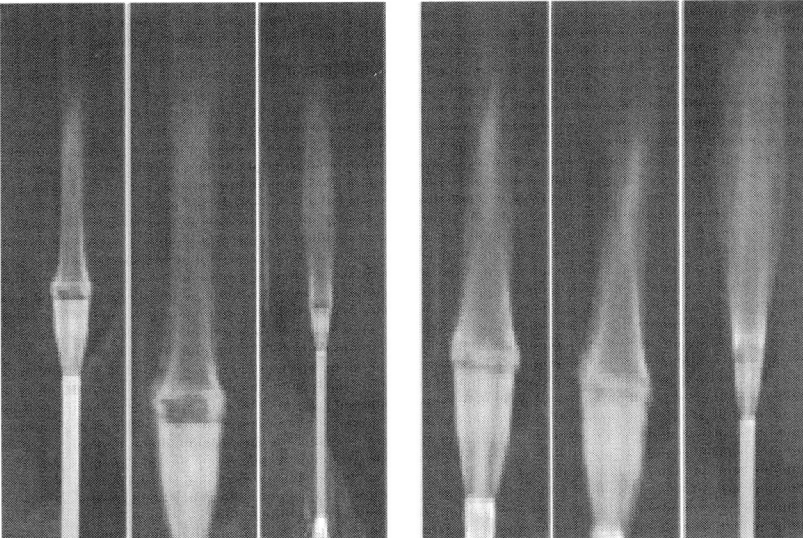

Figure 5: Flame of Bunsen burner, with grid type flame stabilizer, on natural gas (left) and dimethylether (right) [14].

Table 3: Chemical composition of syngas and lower heating values resulted from biomass gasification [15]

Syngas chemical composition [%]							Lower heating value [MJ/Nm3]
Name	CO	H2	CH4	CnH2n	CO2	N2	
Dry oak	18.3	16.9	2.8	0.5	16.0	-	5.422
Dry beech	19.4	17.5	2.6	0.6	15.0	49.3	5.526
Dry fir	15.1	19.1	1.6	0.9	15.8	57.1	4.053
Wood coals	31.2	6.3	2.9	-	2.5	57.1	5.702

Table 4: Chemical composition of syngas and lower heating values resulted from different methods of gasification [12]

CO [%]	H2 [%]	CH4 [%]	N2 [%]	H2O [%]	CO2 [%]	LHV [MJ/Nm3]	Observations
16	6	4	56	18	-	4.1	Air gasification
16	6	4	56	15	3	4.1	Air gasification
40	13	15	3	-	29	11.826	Oxygen gasification

Solving the fuels interchangeability issue for gas turbine cogeneration groups, by developing high level combustion technologies for alternative fuels, particularly hydrogen, will have a major impact on system efficiency and environment.

Fuels Interchangeability and Validation Criteria

Interchangeability in gas turbine cogeneration groups represents the capability to replace a gas fuel with another without affecting the application or the installation burning the gas fuel. The used gas fuels consist in mixtures of combustible gases (methane and other light hydrocarbons, hydrogen, carbon monoxide) and inert gas (mostly nitrogen, carbon dioxide, water vapor). Depending on the combustible gases ratio (usually methane), the gas fuels can have high or low heating value. Density and temperature of the used fuel, as well as the environmental temperature, can affect the performances and lifespan of the equipments in the cogeneration group. According to these influence factors, the most important parameter for characterizing the interchangeability is the Wobbe index (named after engineer and mathematician John Wobbe), defined as ratio between the lower heating value (LHV) and the sqare root of density of the fuel, relative to air density (d_{rel}):

$$Wo = LHV/(d_{rel})^{0.5} \tag{1}$$

$$d_{rel} = \rho_{comb}/\rho_{air} \tag{2}$$

Therefore, two gas fuels, with different chemical compositions but the same Wobbe index, are interchangeable and the heat delivered to the equipment is equivalent for the same fuel pressure. Table 5 gives the values of Wobbe index for several gas fuels. In order to consider the temperature of the fuel, the Wobbe index can be corrected with the temperature. According to [17], two fuels are interchageable if they respect:

$$\frac{\Delta p_2}{\Delta p_1} = \left(\frac{Wo_1}{Wo_2}\right)^2 \left(\frac{A_1}{A_2}\right)^2 \tag{3}$$

where Δp_1 and Δp_2 represent the overpressure of fuel 1, respectivelly 2, Wo_1 and Wo_2 – Wobbe indexes of fuel 1, respectively 2, A_1 and A_2 – injection nozzle area for the two fuels.

Table 5: Wobbe index for various gases [2, 13, 14]

No.	Gas name	Wobbe index [(MJ/Nm3]
1	Natural gas	48.554
2	Liquefied petroleum gas	79.993
3	Methane	47.947
4	Ethane	62.513
5	Propane	74.584
6	Carbon monoxide	12.812
7	Biogas	27.3
8	Dimethylether	47.422
9	Hydrogen	38.3

Therefore, the validation criteria for replacing a fuel with an equivalent one are given by: autoignition temperature, flame temperature (with higher influence on NO_x formation), flame velocity, flashback, efficiency, NO_x and CO emissions, flue gas dew point, etc. Autoignition temperature of gas fuel in mixture with air is the temperature on which the instantaneous and explosive autoignition occurs, without the existence of an incandescent source of ignition. The turbulent flame is generally less stable than the laminar flame, the instability in flame front break-up field being emphasized by the increase in tube diameter. Free swirl turbulent flames are more prone to flame front break-up than the laminar ones due to the higer periferal jet velocity. For turbulence angles greater than 30°, the stability area is achieved on the contour of the burner only for rich mixtures [18]. In areas with poor mixture, due to the decrease in velocity, the backflow can occur without flame attachment on the burner edge. The velocity distribution in the swirl flow determines the stabilisation of the flame as a central suspended one. Components with rapid burning, such as hydrogen, accelerate the flame velocity with a tendency to backflow or extinguishment. The backflow tendency of the flames is proportional with the ignition velocity of the fuel gas, a high velocity leading to a high effect. It is also dependent of the primary air proportion and the components with reduced burning velocity can lead to flame front break-up. In order to consider these factors, an empiric relation has been established for the flame front break-up index at interchangeability I_{ret} [19]:

$$I_{ret} = \frac{k_i f_i}{k_b f_b} \left(\frac{LHV_i}{LHV_b} \right)^{0.5}$$

(4)

where: k – constant concerning the flame front break-up limit; f – factor concerning primary air; LHV – lower heating value; b and i – indexes regarding the control fuel, respectively the replacement fuel. A particular issue is raised by the fuels with reduced heating value. Therefore, the landfill gas contains over 40 % CO_2, requiring a suitable fuel feeding in order to achieve combustion. The fuels with reduced heating value have a small range of flammability requiring, at partial loads or transient operating regimes, the utilization of a supplementary fuel (such as propane). The mass flow rates necessary for gas turbine operation on reduced heating value gas fuels are high (neglecting the water or steam injection in the gas turbine) compared with the operation on natural gas, fact that modifies the compressor's operating characteristic [20]. From biomass gasification with air, it is obtained syngas with LHV of 4-6 MJ/Nm³, and from the gasification with steam or oxygen (see table 4) LHV of 9-13MJ/Nm³. An alternative for increasing lower heating value is the mixing with natural gas. Therefore, if the landfill gas has a LHV of 17-20 MJ/Nm³, an equivalent lower heating value can be obtained by mixing 60 % gas with reduced heating value with 40 % CH_4, with respect to the composition described in [21].

Converting the Aviation Gas Turbines from Liquid to Gas Fuels Operation

The complexity of thermo-gas-dynamic processes defining the gas turbine operation in a cogeneration group require theoretical and experimental research activities on gas turbines in order to accomplish the conversion from liquid to gas fuels operation. For the gas turbines on market, in exploitation, the exploitation and maintenance technical specifications are generally known, being provided by the producer. When the object of the research is an existing gas turbine lacking the technical documentation which completely define the contructive solution, the issue must be approached through activities of experimentation, measurements, CAD 3D modelling, numerical simulation in CFD environment, constructive modifications and renewed experimentation in order to validate the constructive solutions, permanently aiming the performances correlated with the maximum effectiveness (thrust, power), minimum specific fuel consumption, maximum efficiency, versatility on fuel conversion, maximum availability, minimum operation and maintenance costs.

General Criteria – researches Concerning the Modifications on a Gas Turbine for Gas Fuel Operation

The basic procedure for an aeroderivative gas turbine is to keep the rotor assembly, compressor – turbine, which is the „heart" of the gas turbine, form the aviation gas turbine and to redesign the combustion chamber in order to operate on a different fuel than the kerosene. Therefore, for the basic gas turbine in the turboshaft category, at least the combustion chamber must be designed for gas fuels operation. The shaft of the power turbine is mechanicaly connected to a driven load, mechanical work consumer, depending on the application involving the aero-derivative gas turbine (electric generator, compressor, pump, etc.). The command and automatic control system of the aero-derivative gas turbine are designed depending on the application. The bearings can be redesigned, achieving a conversion from rolling bearings to slide bearings. For the basic turboprop (destined for propeller aircrafts), at least the combustion chamber and the reducing gear box and/or the gas generator's turbine must be redesigned, depending if the turboprop does or does not include free turbine. Usually, only the gas generator is used, eliminating the gear box. The issues concerning the automatic control system and the bearings are identical to those of the turboshaft. For the basic turbojet (simple flow jet predominantely for military aircrafts) the redesigning of the combustion chamber and the designing of a power turbine gas-dynamicaly connected to the gas generator are necessary [2]. The issues concerning the automatic control system and the bearings are also identical to those of the turboshaft. Regarding the combustion chamber, is desired to constructively alterate it as little as possible, maybe only in terms of injection system. Due to the fact that the rest of the parameters characterizing the operating process remain unchanged, those regarding zero velocity and ground conditions of the basic gas turbine, the operation of the combustion chamber can be considered as in terms of gas-dynamic similarity. A first problem that must be studied when replacing the fuel is maintaining the combustion efficiency. A second one concerns the maintaining of constructive-functional temperature distribution (on the walls of the firing tube, in the outlet area of the combustion chamber and inlet area of the turbine). On the background of the assembly gas-dyanmic characteristics, the unevenness of the temperatures field on the outlet of the combustion chamber (temperature map) is determined by the geometric characteristics of the dilution area (diameter, length, number and area of holes, etc.) and the characteristics of fuel feeding in the primary area (atomization, jet angles, fuel specifications, etc.). The global temperature map is defined by equation (5) and the radial unevenness for the rotor bladed area is given by equation (6) [3]:

$$\theta_m = \left(T^*_{\max} - T^*_3\right) / \left(T^*_3 - T^*_2\right)$$

(5)

Radial unevenness for the rotor bladed area express the manner of operation on the turbine blades:

$$\theta_r = \left(T^*_{\max r} - T^*_3\right) / \left(T^*_3 - T^*_2\right)$$

(6)

In equations (5) and (6) the significance of symbols is: $T^*_{\max}$ - maximum temperature peak; T^*_3 - average temperature in the outlet section of the combustion chamber; T^*_2 - average temperature in the outlet section of the compressor; $T^*_{\max r}$ - maximum average radial temperature, circumferential arithmetic mean on the entire section. Normal values for θ_m, depending on the gas turbine, are in the 20-25 % range, with reported values of 35 %. In direct connection with the temperature map on the walls of the firing tube, the equivalent stress of the material must be considered when replacing a fuel with another. In the case of the AI 24 gas turbine modification for operation on gas fuels in the cogeneration group, a difference of 15 % has been reported in the temperature map, considering the flattening of the temperature peaks when passing through the turbine [22]. The adopted solution has been the generalization of the results obtained by National Research and Development Institute for Gas Turbines COMOTI Bucharest for the AI 20 GM (figure 1) and MK 701 gas generators. In order to achieve the AI 20 GM gas turbine on natural gas (derived from AI 20 on liquid fuel) the adopted constructive solution has been the modification of the injection system, without altering the firing tube (figure 6). The researches for this transformation have been based on test bench experiments with liquid fuel (in low pressure similitude conditions). In order to reach the functional optimum on natural gas, several injection nozzles have been designed and experimented, according to table 6 [3].

Table 6: Configuration of the experimental injection nozzles (see figure 6), for AI 20 GM on natural gas [3]

Nozzle no.	10 Ø3 holes at a 2 angle	Diameter of central hole [mm]
1	900	3
2	700	without central hole
3	800	without central hole
4	700	3
5	800	3
6	1000	without central hole

Only nozzles with 10 holes of the same diameter have been experimented in order to ensure velocity, penetration and safety in operation. The central hole afects the stability of the combustion process, increases the flame radiation and the temperature on the walls of the firing tube. The tie criterion for various injection nozzles for natural gas has been the temperature of the blade on hub. It has been noted that nozzle no. 3 leads to low frequency vibrations in a large range of operating regimes, functionally inadmissible. When operating on natural gas, the combustion efficiency increases with the operating regime, the process being unaffected by the vaporization, but only by the mixing. Following the experimentation, nozzle no. 2 has been selected (with 10 Ø3 holes at $2\alpha=70^0$, without central hole). For all experimentation regimes, the circumferential temperature map values did not pass 18 %. The same manner of minimum configuration modifications has been applied for the rest of the gas turbines transformed for operating on natural gas (TURMO, MK 701, etc.). Therefore, the firing tube and the combustion chamber case have been kept unmodified for all gas turbines, only redesigning the injection system. Satisfying results have been obtained for the experimentation of TURMO: good stability, but in a more limited range compared with other gas turbines (due to the dependency on the mixing process); temperature map values of 22 % (for the aimed 20 %). For MK 701, the values on the temperature map have reached max. 20 %. A particular problem is considered when the aim is the integration of the gas turbine, modified for operating on natural gas, with an existing boiler. The heat recovery steam generator can be derived form an energy steam boiler, a technological steam boiler or a hot (warm) water boiler. The integration analysis for an aeroderivative gas turbine with a hot water boiler shows that the temperature of the burned gases on the stack must be in the usual value range and the pressure loss at the passing through the modified boiler (in the cogeneration group) must be lower than the pressure loss on the initial boiler [23]. The modifications necessary for operating the gas turbine on gas fuels with reduced lower heating value, compared with the operation on natural gas, are slightly more complex. Therefore, Mitsubishi Heavy Industries Ltd., with extensive experience in manufacturing gas turbines on BFG (Blast Furnace Gas), considerd the heating value of the gas fuel as the key factor in the modifications scheduled for the gas turbine [24]. Depending on the actual application, more modifications can be operated on the gas turbine, compared with the ones in table 7.

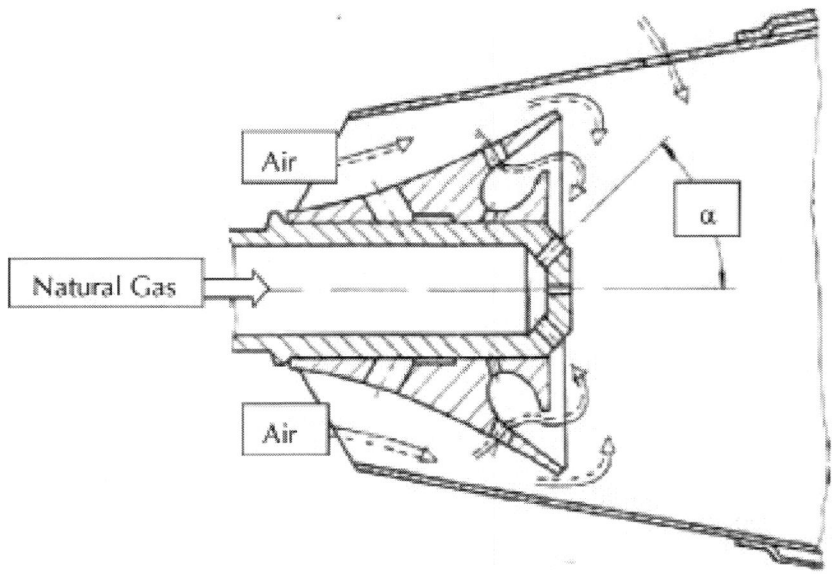

Figure 6: Modification of the injection system for AI 20 GM gas turbine [3].

Table 7: Necessary modifications for a gas turbine, depending on lower heating value of the fuel [24]

Lower heating value [MJ/ Nm3]	20.95-41.9 (High)	35.61 (Natural gas)	8.38-29.33 (Medium)	2.51-8.38 (Low)
Air compressor	Standard	Standard	Standard	Modification
Combustor	Standard (Minor mod.)	Standard	Standard (Minor mod.)	Modification
Turbine	Standard	Standard	Standard	Standard
Fuel system	Standard (Minor mod.)	Standard	Standard (Minor mod.)	Modification

Converting a Gas Turbine from Liquid to Gas Fuel Operation for Landfill Gas Valorisation

Converting the gas turbine from liquid fuel to gas fuel operation in order to achieve the valorisation of the landfill gas has known two main steps, respecting the principles in chapter 2.3: converting the TV2-117A gas turbine from operating on liquid fuel (kerosene) to gas fuel (natural gas), resulting the TA2 gas turbine; converting the TA2 from operating on natural gas to operating on landfill gas, resulting TA2 bio. In order to achieve these results, numerous numerical simulations in CFD environment and tests have been used for validating the adopted solutions.

Numerical Simulation, Experimental Activity, Methods and Equipments

Numerical simulation on the TV2-117A gas turbine (figure 7, left) on kerosene has been made in order to obtain a reference model for the gas turbine conversion on gas fuels, particularly landfill gas. An eighth of the geometric model, corresponding to one injection nozzle, has been used in simulations considering the combustion chamber simetry. The boundary conditions have been provided by the producer in the technical specifications for three operating regimes: take-off, nominal and cruise (with the corresponding temperatures of 1123, 1063 and 1023 K). For simulating the combustion process in the TA2 bio gas turbine, the used fuel has been a synthetic landfill gas with equal volume proportions of methane (CH_4) and carbon dioxid (CO_2). The real landfill gas contains other chemical species, in small proportions, which have been considered impurities and have not been taken into account. The numerical simulations have been made on the TA2 with modified injection system, particularly on the injection nozzles level (figure 8). The modelling of the injection nozzles has been achieved starting from the geometry of the natural gas nozzles. Only the injector's outer body have been kept from the liquid operating gas turbine, eliminating all elements related to the atomization system of the liquid fuel. Related to the initial configuration of the injector, only the diameter of the secondary channel and the configuration of the connection with the injection nozzle have been kept unmodified.

Figure 7: TV2-117A gas turbine (left) with detailed combustion chamber area (right).

The numerical simulations for the modified injector (figure 8) have taken into consideration the variation of the injection pressure (7.65 - 8.5 bar), of the injection angle β (70 - 85⁰) and the position related to the injector's body L (1 - 5 mm). Following the numerical simulations, the optimum configuration has been selected and the eight injectors have been manufactured along with the injection ramp (figure 10, right), consisting in a circular pipe connected to each injector. The configurations of the injectors for liquid fuel and landfill gas are given in figure 9. The elements eliminated from the initial configuration are the following: the liquid fuel feeding system; the liquid fuel automatic control system; the command system for the actuators controlling the guide vanes and the first three statoric stages of the compressor; the deicing system. The experimentation of TA2 bio has been made in the experimental facility of National Research and Development Institute for Gas Turbines COMOTI Bucharest (figure 10) in the following configuration: TA2 bio gas turbine installed on test bench; test cell lubricating system and fuel feeding system for the gas turbine; exhaust system for the burned gases; monitoring system for acquiring functional parameters. In figure 10 (right) is a pipe ramp ring, yellow color, for gas fuel supply.

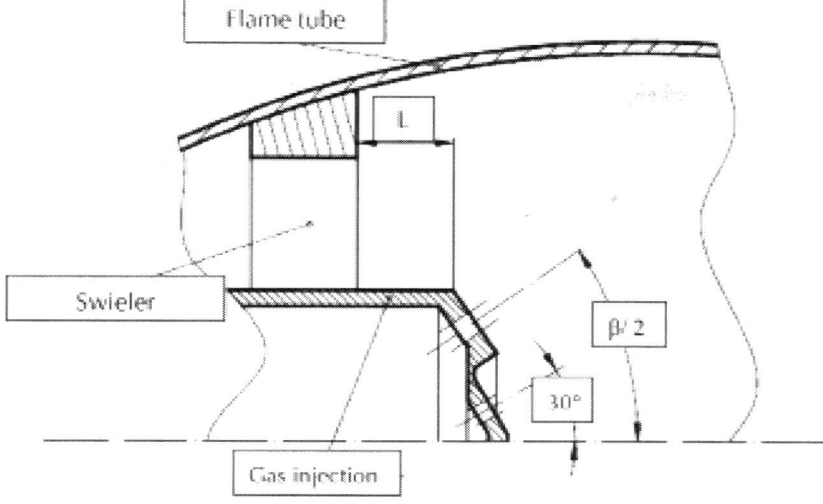

Figure 8: Injection nozzle configuration for landfill gas [2].

Figure 9: Injectors for liquid fuel for TV2-117A (left) and landfill gas for TA2 bio (right) [2].

A series of experimentations have been made, the simulated landfill gas being obtained by mixing natural gas with carbon dioxid (provided from tanks). The measurements have been made with the equipments of the test facilities. A ramp of 17 double thermocouples located at the outlet of the combustion chamber, with measuring points at one third and two thirds of the outer firing tube circumference allow the measurement of the T_{ex} and T_{in} temperatures on two concentric rings (figure 11 right).

Figure 10: TA2 gas turbine (left) and TA2 bio gas turbine in the test cell (centre, right).

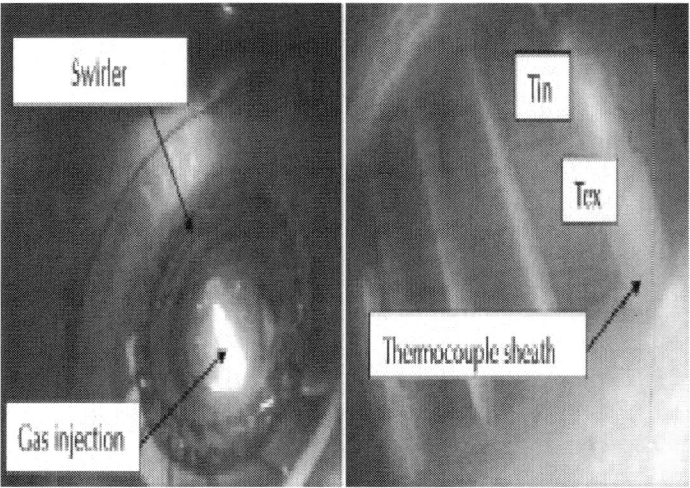

Figure 11: Boroscoping images of the gas injection nozzles – natural gas (left) and the thermocouples (right).

Results and Discussion

The numeric simulations on kerosene [2, 5] have shown that, for the reduced operating regimes, the flame reaches in high degree the area between two adjacent injectors. Table 8 presents the numerical results for landfill gas combustion in terms of methane mass fraction, illustrating the jet shape, and burned gases temperature in the oultlet section of the combustion chamber. Analysis of data in table 8, with respect to temperature maps, aiming to obtain a compact jet in order to protect the walls of the firing tube, have helped selecting the geometric configuration of the injection nozzle: $\beta = 70^0$ and L= 3 mm, used for designing the functional model experimented on TA2 bio, for a mixture of natural gas and carbon dioxide. The experiments have been developed in several series, figure 12 presenting one of the models of variation for the components of the synthetic landfill gas mixture. The experimental results have been synthetized in figures 12 and 13. Figure 13 presents the numerical and experimental results for the outlet section of the combustion chamber.

Table 8: Numerical results for landfill gas combustion simulation [2]

Parameters			Results	
L [mm]	[0]	p [bar]	Fuel injection jet	Temperature on combustion chamber outlet
1	70	7.65		
3	70	7.65		

5	70	7.65		
3	80	7.65		
3	85	8.50		

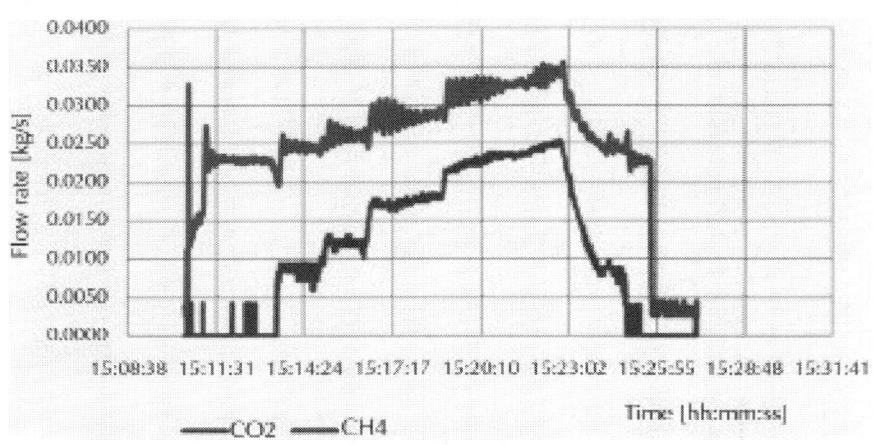

Figure 12: Variation of the mass flow rates of carbon dioxide (CO_2) and natural gas (CH_4) injected in the combustion chamber [2].

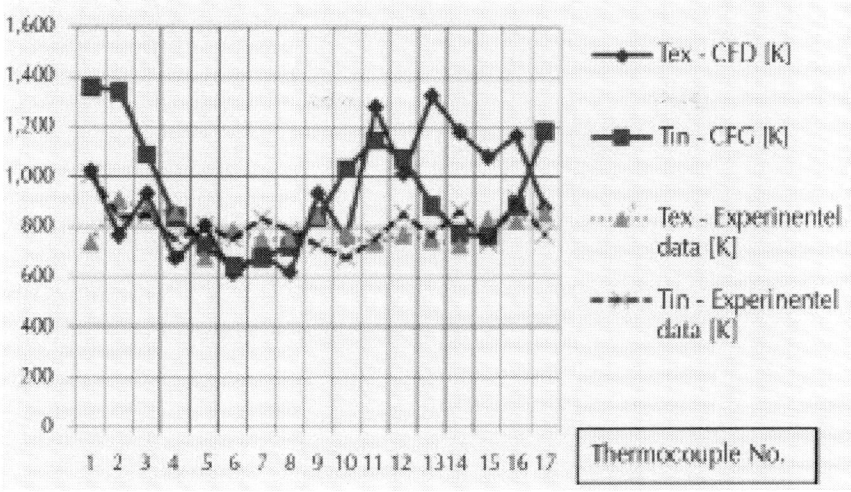

Figure 13: Comparison between the numerical and experimental temperature.

The experimentations have proved a stable operation of the TA2 bio gas turbine on different operating regimes, mainly defined by the mass flow rate and the ratio between the mass flow rate of the natural gas and carbon dioxid. Figure 13, particularly the central area, shows a concordance of the numerical and experimental data, proving that modification of gas turbines operating on alternative gas fuels can be made based on numerical simulations in CFD environment. The model of a cogeneration plant for electric and thermal energy is illustrated in figure 14.

Figure 14: Model of an aeroderivative gas turbine cogeneration plant operating on natural gas and landfill gas.

FLEXIBILITY OF GAS TURBINE COGENERATION GROUPS AND EMISSIONS REDUCTION – FUTURE RESEARCHES

Gas turbine cogeneration groups, alone or in combination with fuel cells, can play an importan role in the general assembly of energy production and emissions reduction. The NO_x reduction must be regarded considering the ensurance of cogeneration group performances in a flexible manner, optimization being possible for a fuel [25]. A higher efficiency implies the optimization of the entire cogeneration plant (gas turbine, afterburning, heat recovery steam generator, etc.). The efficiency must be maintained for partial loads (even below 50 %) or for environmental conditions modification. Starting from 2002, Siemens has taken into consideration the flexibility, eliminating the high pressure barrel of the heat recovery steam generator which requires a long process to reach a certain temperature (in order to avoid the occurence of thermal tensions). Regarding the flexibility, the efficiency and the emissions reduction in gas turbine cogeneration groups, important steps have been made: reduced NO_x burners have been introduced in applications; the lifecycle has been analyzed for efficiency increase; the period between maintenence controls has been extended and the conversion from one fuel to another for multi-fuel engines has improved [7]. The factors determining the formation of pollutant agents exhausted along with the burned gases from the gas turbines are [26]: temperature and air excess coefficient in primary area; homogenization of the process in primary area; residence time of the products; "freezing" characteristic of the reaction near the firing tube, etc. For NO_x reduction, the temperature in the area of the combustion reaction and the areas of maximum temperature and the air jets distribution (stage combustion) need to be reduced. The final configuration of the combustion chamber of a gas turbine is a compromise between the NO_x level, performance and flexibility. Global reduction of the emissions leads to compromises between the emission levels of different components and the assembly characteristics of the combustion chamber (pressure losses, stability and ignition limits, etc.). New concepts must be promoted in order to solve this issue. The usual methods are represented by the water or steam injection in the combustion chamber of the gas turbine, leading to [12]: reduction of NO_x up to 25 ppm (for a 15 % O_2 volume participation in dry burned gases); increase in turbine power due to the increase in fluid mass flow rate (which can compensate the effect of increased temperature during summer); increase of flexibility of the installation in exploitation due to the possibility of load variation through steam flow rate variation. However, the high content of vapours in burned gases can lead to: acid corosion occurence (for fuels containing sulphure); increase in thermal stress on the combustion chamber; reduction of the heat recovery level, etc. Numerical simulations on TV2-117A (figure 15) for water injection in the combustion chamber (through duplex injectors, on natural gas) have shown that the water injection in truncated cone shape, at 45°, characterized by a 12 l/min mass flow rate, leads to minimum NO_x concentration in burned gases of 14 ppm. The analysis of combustion products for TA2 (see chapter 2.4), using NASA CEA program [27], has shown a decrease of the average maximum temperature. The composition of the landfill gas has been considered in equal volume proportions of methane and carbon dioxide, while the composition of the syngas has been considered that given by [19]. The calculation algorythm has started from the stoichiometric reaction of each fuel and imposing the operating regime (in terms of average maximum temperature of 1063 K for nominal regime) in order to determine the minimum quantity of air necessary for the reaction. Obtaining the equilibrium reactions has determined the calculation of the air excess coefficients for each fuel at the given regime, for dry operation. Starting from these initial values, water has been introduced in different proportions, up to 23 %. The supplementary quantity of fuel, necessary to reestablish the operating regime of the gas turbine, in terms of temperature (considering the pressure as unaffected), has been calculated in relation to the quantity of water. The general combustion reactions for each fuel, for the water injection case, for the nominal operating regime, are given by equation (7)for landfill gas and equation (8) for syngas:

$$b \cdot (CH_4 + CO_2) + 2 \cdot \lambda \cdot (O_2 + 3.76\ N_2) + a \cdot 2 \cdot \lambda \cdot H_2O \rightarrow w\ H_2O + x\ CO_2 + y\ N_2 + z\ O_2$$

$$\text{(7)}$$

$$b \cdot (0.25\ CO + 0.09\ CO_2 + 0.12\ H_2 + 0.52\ N_2 + 0.02\ CH4) + 0.225 \cdot \lambda \cdot (O_2 + 3.76\ N_2) + a \cdot 0.225 \cdot \lambda \cdot H_2O \rightarrow w\ H_2O + x\ CO_2 + y\ N_2 + z\ O_2$$

$$\text{(8)}$$

There have been tracked the thermodynamic of the system and the concentrations of the reaction products, focusing on carbon monoxid (CO) and nitrogen oxides (NO_x). In these conditions, for the two regimes, the calculations have been made up to a injected water coefficient (noted „a") in oxidant of maximum 2, equivalent to 23 % water in oxidant. The maximum proportion of water in oxidant has been limited by the concentration of oxygen resulted from the combustion, minimum 11 %, necessary for the afterburning process. For the nominal operating regime and approximately 15 % water for landfill gas and 12.5 % for syngas, the gas turbine reaches the minimum limit of oxygen.

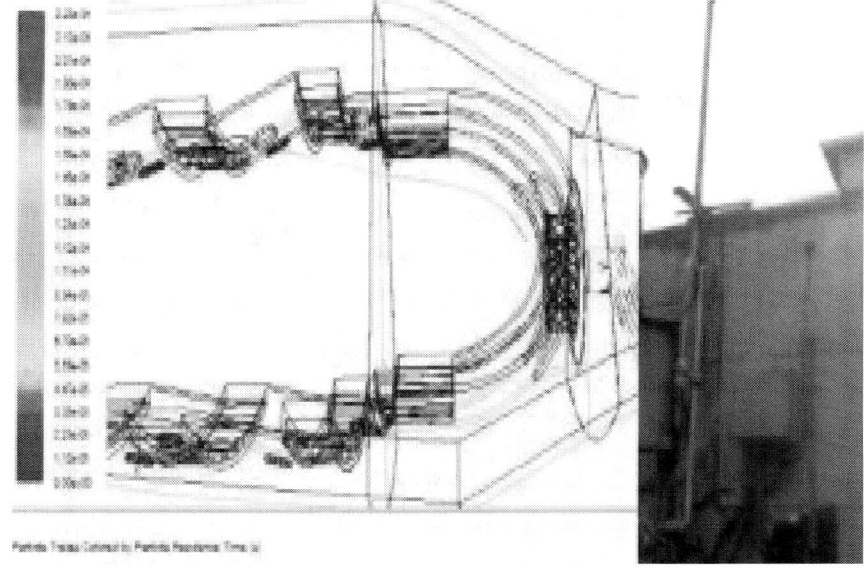

Figure 15: Numerical simulation of water injection in the combustion chamber of TA2 (left) and atomization tests with the duplex injector (right).

Figure 16 shows the variation of NO_x for the two fuels (landfill gas and syngas) for the nominal regime, depending on the injected water proportion. The results of the calculations illustrate that the use of afterburning along with the operation of the TA2 gas turbine, with water injection, for the good operation of the system, the NO_x produced by the gas turbine at 1063 K can only be reduced to 40 ppm for landfill gas and 38.5 ppm for syngas. The oxygen injected in the air can lead to nitrogen oxides reduction and combustion enhancement resulting [28]: reduction of ignition temperature; increase in flamability limit; increase in adiabatic temperature of the flame; increase in process stability and control; reduction of low heating value fuels consumption, etc. The adiabatic temperature of the flame increases with approximately 50 °C for 1 % increase in oxygen concentration. The volume of burned gases decreases with 12 % for the combustion of natural gas in 3 % oxygen enriched air [29]. Reduction of pollution through combustion in oxygen enriched environment can be used in afterburning installations (for primary or secondary air). Combustion in oxygen enriched environment can increase the efficiency and the flexibility of the cogeneration plant. When adding hydrogen to a gas fuel, there are affected the stability of the flame, the efficiency of the combustion and the emissions. Flame velocity for hydrogen combustion in air, in stoichiometric conditions, reaches 200 cm/s compared to the combustion of methane in air, for which the velocity is approximately 40 cm/s [29]. Adding hydrogen to the gas fuel of the gas turbine or afterburning installation can lead to CO and NO_x emissions reduction.

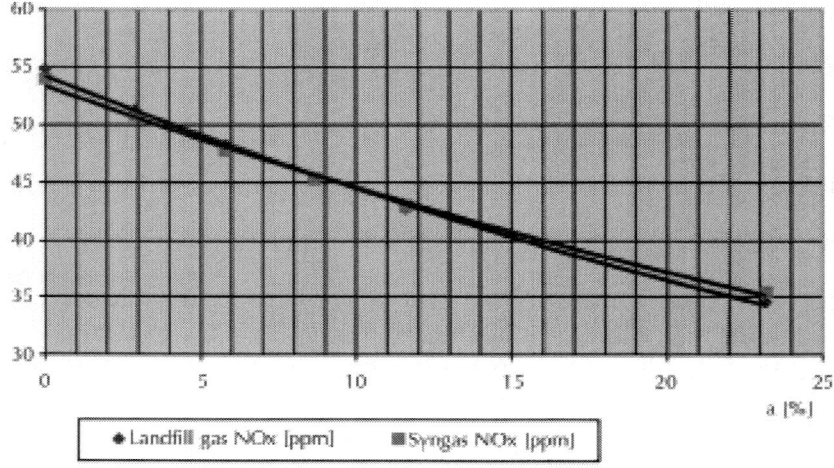

Figure 16: Variation of NO_x concentration for the two fuels, at 1063 K, depending on water proportion in oxidant (a).

Afterburning Installation as Interface between Gas Turbine and Heat Recovery Steam Generator

The burned gases flow when exiting the gas turbine is turbulent and unevenly distributed in transversal section. Therefore, backflow can occur in the transversal section of the recovery boiler. The unevenness of the flow and the variation in burned gases composition affects the operation of the afterburning. Therefore, the afterburning is influenced in terms of efficiency, emissions, flame stability, as well as corrosion of the elements subjected to the action of burned gases. For a good design of the inlet section in the recovery boiler it must be generally considered the following factors [30]: geometry and direction of the gas turbine exhaust; size of heat exchange surfaces; location of the afterburning burner; mass flow rate and average velocity of burned gases exiting the gas turbine; local velocities near the walls and on the first heat exchange surface. The gas turbine exhaust is generally not directly connected with the recovery boiler. After exiting the gas turbine (the case of 2xST 18 Cogeneration Plant at Suplacu de Barcau), the burned gases pass through a silencer, a by-pass assembly, a transom for the connection with the burner and then the afterburning chamber [8]. The gases flow must be parallel with the axis of the burner's connector (perpendicular to the burner plane). A uniform distribution of the flow in the transversal section ensures a good operation of the heat recovery steam generator, particularly regarding the superheater. Therefore, the necessary premises are created for ensuring low emissions on the cogeneration group. If the burned gases or the air are uneven distributed, significant variation of the temperatures downstream the burner can occur. Velocity variation in the transversal section, upstream the burner, must not exceed, on 90 % of the burner's section, ± 15 % of the average velocity measured on the entire transversal section. In reality, the burned gases temperature downstream the burner will never be perfectly uniform. Even for a perfect flow distribution of the turbine gases, upstream the burner, the temperature in the area of each burner module will be higher than the temperature between the modules. Therefore, the infrared analysis of the channel connecting the gas turbine and the afterburning installation (silencer – by-pass assembly – connecting transom), at 2xST 18 Plant, has shown unevenness in temperature distribution (figure 17). Considering these phenomena, the afterburning installation can compensate, in good conditions, the mass flow decrease in burned gases produced by the gas turbine at partial loads, keeping a corresponding load on the heat recovery steam generator. In case of turbine stopping, the heat recovery steam generator with the fresh air afterburning is able to keep the steam production at a certain level.

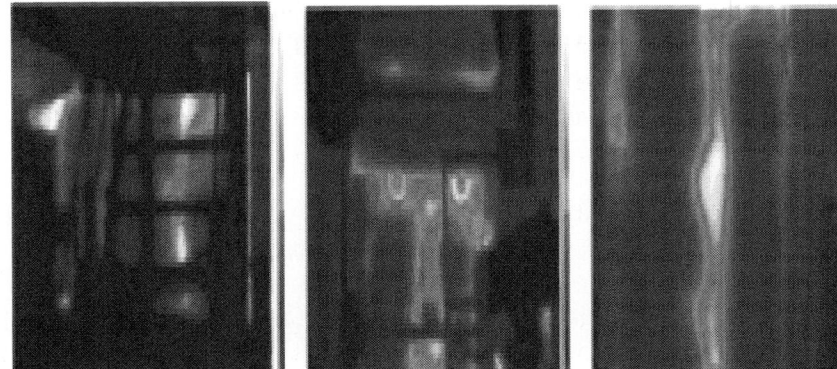

Figure 17: Temperature isotherms, in infrared, in the channel connecting the gas turbine and the afterburning installation (silencer – by-pass assembly – connecting transom).

Future Research

Future research is part of the general context of increasing the flexibility of gas turbine cogeneration groups, the efficiency and reducing the emissions using numerical simulations in CFD environment and experimentations related to: utilization of alternative fuels in gas turbines and afterburning installations, injection of fluids in the cogeneration line in order to reduce the emissions, integrating the gas turbine with fuel cells, etc.

CONCLUSIONS

Along with the flexibility to alternative fuels feeding, the flexibility of a gas turbine cogeneration plant assumes the accomplishment of several requirements: capability of fast start; capability to pass easily from full load to partial loads and back; maintaining the efficiency at full load and partial loads; maintaining the emission to a low level even when operating on partial loads. Using aeroderivative gas turbines in the cogeneration field has allowed the scientific and technologic knowledge transfer utilization (design concepts, materials, technologies, etc.), which ensures a high degree of energy, from aviation to ground applications. The experience of National Research and Development Institute for Gas Turbines COMOTI Bucharest, in the field of aeroderivative gas turbines (AI 20 GM, TURMO, MK 701, etc.) has allowed the conversion of a gas turbine from liquid fuel to landfill gas, for cogeneration, in stable operating conditions.

REFERENCES

1. Cenusa V., Benelmir R., Feidt M., Badea A. On gas turbines and combined cycles. http://www.ati2001.unina.it/newpdf/Sessioni/Macchine/Impianti/03-Cenusa-Benelmir-Feidt-Badea.pdf (accessed June 5, 2012).

2. Petcu R. Contributii teoretice si experimentale la utilizarea gazului de depozit ca sursa de energie. Teza de doctorat - Decizie Senat nr. 100/12.02.2010. Universitatea Politehnica Bucuresti; 2010

3. Carlanescu C. Contributii la problema selectarii si modificarii motoarelor de aviatie pentru utilizarea in scopuri industriale. Teza de doctorat. Universitatea Gheorghe Asachi Iasi; 1994

4. Energy and Environmental Analysis. Technology Characterization: Gas Turbines. http://www.epa.gov/chp/documents/catalog_chptech_gas_turbines.pdf (accessed June 6, 2012).

5. Barbu E., Vilag V., Popescu J., Ionescu S., Ionescu A., Petcu R., Cuciumita C., Cretu M., Vilcu C., Prisecaru T. Afterburning Installation Integration into a Cogeneration Power Plant with Gas Turbine by Numerical and Experimental Analysis. In: Ernesto Benini (ed.), Advances in Gas Turbine Technology. Rijeka: InTech; 2011. p. 139-164. Available from http://www.intechopen.com/articles/show/title/afterburning-installation-integration-into-a-cogeneration-power-plant-with-gas-turbine-by-numerical-(accessed June 6, 2012).

6. Stationary Sources Branch. Stationary Gas Turbines - 40 CFR Part 60. http://www.cdphe.state.co.us/ap/down/statgas.pdf (accessed June 6, 2012).

7. Breeze P. Efficiency versus flexibility: Advances in gas turbine technology. PEI 01/04/2011. http://www.powerengineeringint.com/articles/print/volume-19/issue-3/gas-steam-turbine-directory/efficiency-versus-flexibility-advances-in-gas-turbine-technology.html (accessed May 31, 2012).

8. Barbu E., Ionescu S., Vilag V., Vilcu C., Popescu J., Ionescu A., Petcu R., Prisecaru T., Pop E., Toma T. Integrated analysis of afterburning in a gas turbine cogenerative power plant on gaseous fuel, WSEAS Transaction on Environment and Development, 2010; 6(6) p. 405-416. http://www.wseas.us/e-library/transactions/environment/2010/89-806.pdf (accessed June 5, 2012).

9. Jones R., Goldmeer J., Monetti B. Addressing gas turbine fuel flexibility. GE Energy. http://www.ge.com/cn/energy/solutions/s1/GE%20Gas%20Turbine%20Fuel%20Flexibility.pdf (accessed June 6, 2012).

10. Pimsner V., Vasilescu C., Radulescu G. Energetica turbomotoarelor cu ardere interna. Bucuresti, Editura Academiei RSR, 1964

11. Marco Antonio Rosa do Nascimento and Eraldo Cruz dos Santos. Biofuel and Gas Turbine Engines, Advances in Gas Turbine Technology. In: Ernesto Benini (ed.), Advances in Gas Turbine Technology. Rijeka: InTech; 2011. p. 116-138. InTech, Available from: http://www.intechopen.com/books/advances-in-gas-turbine-technology/biofuel-and-gas-turbine-engines (accessed June 6, 2012).

12. Oprea I. Posibilitati de utilizare a gazelor provenite din biomasa in instalatii de turbine cu gaze. ETCN-2005, 30 iunie-1 iulie 2005, Bucuresti, p. 135-139

13. Jensen J., Jensen A. Biogas and natural gas, fuel mixture for the future. 1st World Conference and Exihibition on Biomass and Energy, 2000, Sevilla. Available from http://www.dgc.eu/pdf/Sevilla2000.pdf (accessed June 11, 2012).

14. Panoiu P., Marinescu C., Panoiu N., Oroianu I., Mihaescu L. Posibilitati de utilizare a dimetileterului in scopuri energetice. http://caz.mecen.pub.ro/panoiu.pdf (accessed June 11, 2012).

15. Calin L., Jadaneant M., Romanek A. Gazeificarea biomasei lemnoase. Curierul AGIR, 1-2, ianuarie-iunie 2008, p. 87-90

16. Chiesa P., Lozza G., Mazzocchi L. Using hydrogen as gas turbine fuel, Journal of Gas Turbine and Power, January 2005, vol. 127 73-80 http://www.netl.doe.gov/technologies/coalpower/turbines/refshelf/igcc-h2-sygas/Using%20H2%20as%20a%20GT%20Fuel.pdf (accessed June 12, 2012).

17. Ionel I., Ungureanu C., Popescu F. Analiza nivelului de emisii poluante prin schimbarea combustibilui la cuptoarele de tratament termic. http://www.tehnicainstalatiilor.ro/articole/images/nr12_76-82.pdf (accessed June 14, 2012).

18. Antonescu N., Polizu R., Muntean V., Popescu M. Valorificarea energetica a deseurilor. Bucuresti. Editura Tehnica; 1988

19. Ionel P., Borcea Fl., Barbu E., Marinescu C., Ciobanu C. Mihaescu L. Utilizarea combustibililor gazosi regenerabili pentru producerea de energie.Bucuresti. Editura Perfect; 2008

20. Rainer K. Gas turbine fuel considerations. http://www.scribd.com/doc/76918626/Gas-Turbine-Fuel-Considerations (accessed June 14, 2012).

21. Fossum M., Beyer R. Co-combustion: Biomass fuel gas and natural gas. http://media.godashboard.com/gti/IEA/ieaCofirNOrep.pdf (accessed June 16, 2012).

22. Ene M., Ion C., Salcianu R. Cercetari de transformare a unei camere de ardere pentru functionare cu gaze naturale. In: TURBO '98, 13-15 iulie 1998, Bucuresti, Romania

23. Zubcu V., Zubcu D., Stanciu D., Homulescu V. Instalatie de cogenerare cu componente recuperate, conditii de compatibilitate. in: TURBO '98, 13-15 iulie 1998, Bucuresti, Romania

24. Komori T., Yamagami N., Hara H. Design for blast furnace gas firing gas turbine. http://www.mnes-usa.com/power/news/sec1/pdf/2004_nov_04b.pdf (accessed June 20, 2012).

25. Richards G., McMillian M., Gemmen R., Rogers W., Cully S. Issues for low-emission, fuel-flexible power systems. Progress in Energy and Combustion Science 2001; 27: p. 141–169.

26. Carlanescu C., Manea I., Ion C., Sterie St Turbomotoare – Fenomenologia producerii si controlul noxelor. Bucuresti: Editura Academiei Tehnice Militare; 1998.

27. Zehe, M.J., Gordon, S. & McBride, B.J. (2002), *CAP: A Computer Code for Generating Tabular Thermodynamic Functions from NASA Lewis Coefficients*, NASA Glenn Research Center, NASA TP—2001-210959-REV1, Cleveland, Ohio, U.S.A., http://www.grc.nasa.gov/WWW/CEAWeb/TP-2001-210959-REV1.pdf (accessed June 26, 2012).

28. Corna N., Bertulessi G. The use of oxygen in biomass and waste-to-energy plants: A flexible and effective tool for emission and process control, Third International Symposium on Energy from Biomass and Waste, 8-11 November 2010, Venice, Italy

29. Drnevich R., Meagher J., Papavassiliou V., Raybold T., Stuttaford P., Switzer L., Rosen L. Low NOx emissions in a fuel flexible gas turbine, Issued August 2004, http://www.netl.doe.gov/technologies/coalpower/turbines/refshelf/reports/41892%20Praxair%20Final%20Report_Low%20NOx%20Fuel%20Flexible%20Gas%20Turbine.pdf (accessed June 26, 2012).

30. Daiber J., Fluid dynamics of the HRSG gas side, Power, March 2005, p. 58-63 http://www.babcockpower.com/pdf/vpi-45.pdf (accessed June 26, 2012).

Chapter 8

Review of the New Combustion Technologies in Modern Gas Turbines

M. Khosravy el_Hossaini[1]

[1]Research Institute of Petroleum Industry, Iran

INTRODUCTION

The combustion chamber is the most critical part of a gas turbine. The chamber had to be designed so that the combustion process to sustain itself in a continuous manner and the temperature of the products is sufficiently below the maximum working temperature in the turbine. In the conventional industrial gas turbine combustion systems, the combustion chamber can be divided into two areas: the primary zone and the secondary zone. The primary zone is where the majority of the fuel combustion takes place. The fuel must be mixed with the correct amount of air so that a stoichiometric mixture is present. In the secondary zone, unburned air is mixed with the combustion products to cool the mixture before it enters the turbine. In some design, there is an intermediate zone where help secondary zone to eliminate the dissociation products and burn-out soot.

The majority of the combustors are developed base on diffusion flames as they are very stable and fuel flexibility option. In a diffusion flame, there will be always stoichiometric regions regardless of overall stoichiometry. The main disadvantage of diffusion-type combustor is the emission as high temperature of the primary zone produced larger than 70 ppm NOx in burning natural gas and more than 100 ppm for liquid fuel [1]. Several techniques have been tried in order to reduce the amount of NOx produced in conventional combustors. In general, it is difficult to reduce NOx emissions while maintaining a high combustion efficiency as there is a tradeoff between NOx production and CO/UHC production.

In some recent installations, the premixed type of combustion has been selected to reduce NOx emissions bellow 10 ppm. Apart from the flame type change, there are some method such as "wet diffusion combustion", FGR[1] - and SCR[2] - . In an example of wet combustion, a nuzzle through which steam is injected is provided in the vicinity of the fuel injector. The level of NOx emission is controlled by the amount of steam. However, there is a limit on the increasing the steam flow rate as cause corresponding considerable CO emission. Furthermore, preparing pure steam in the required injection condition increases operational costs. Nowadays, wet combustion rarely applies due to water consumption and the penalty of reduced efficiency. Post Combustion treatments such as SCR are those which convert NOx compounds to nitrogen or absorb them from flue gas. These methods are relatively inexpensive to install but does not achieve NOx removal levels better than modern gas turbine combustor.

In this chapter, a short introduction of combustion process and then a description of some new pioneer combustor have been presented. As gas turbine manufacturers are looking for continuous operation or stable combustion, satisfactory emission level, minimum pressure loss and durability or life. Hence, the advanced combustor might include all of these criteria, so some of them are selected to discuss in details.

THE COMBUSTION PROCESS

Type of Combustion Chamber

The diffusion and premixed flame are two main type of combustion, which are using in gas turbines. Apart from type of flame, there are two kind of combustor design, annular and tubular. The annular type mostly recommended in the propulsion of aircraft when small cross section and low weight are important parameters. Can or tubular combustors are cheaper and several of them can be adjusted for an industrial engine identically. Although there are different types of combustors, but generally, all combustion chambers have a diffuser, a casing, a liner, a fuel injector and a cooling arrangement. An entire common layout is visualized in figure 1.

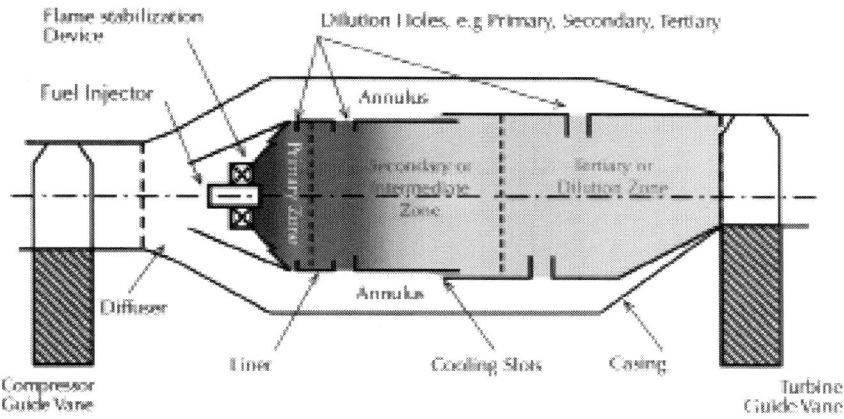

Figure 1: The layout of the combustion chamber.

Flame Stabilization

After the fuel has been injected into the air flow, the flow will enter the flame region. It does this with quite a high velocity, so to make sure the flame isn't blown away; suitable flame stabilization techniques must be applied. First, the high velocity of flow will be responsible for a pressure drop[3] - . Secondly, the flame in the combustion chamber cannot survive if the air has a high velocity. So combustion chambers benefit from diffusers to slow down the air flow. There are two normal kinds of flame stabilizers: bluff-body flame holders and swirlers.

The shape of the bluff–body flame holder affects the flow stability characteristics through the influence on the size and shape of the wake region. Since the flame stabilization depends on size of the zone of recirculation behind the bluff–body, different geometries such as triangular, rectangular, circular and more complex shapes are being use. One of the basic problem of bluff-body flame holders is a considerable effect on pressure loss. Figure 2 shows a high speed image of three flame holders in atmospheric condition.

Figure 2: High speed images of the circular cylinder (top), square cylinder (middle) and V gutter (bottom) at Re = 30,000 and stoichiometric mixture [2].

Flow reversal can be applied in the primary zone. The best way to reverse the flow is to swirl it through using swirlers. The two most important types of swirlers are axial and radial. The advantage of flow reversal is that the flow speed varies a lot. So there will be a point at which the airflow velocity matches the flame speed where a flame could be stabilized. The degree of swirl in the flow is quantified by the dimensionless parameter, Sn known as the swirl number which is defined as:

$$Sn = G\theta Gxr \tag{1}$$

Where:

$$G_\theta = \int_0^\infty (\rho u w + \overline{\rho u' w'}) r^2 dr$$

$$G_x = \int_0^\infty (\rho u^2 + \overline{\rho u^2} + (p - p_\infty)) r \, dr$$

As this equation requires velocity and pressure profile of fluid, researchers proposed various expressions for calculating the swirl number. Indeed, the swirl number is a non-dimensional number representing the ratio of axial flux of angular momentum to the axial flux of axial momentum times the equivalent nozzle radius [3]. Tangential entry, guided vanes and direct rotation are three principal methods for generating swirl flow.

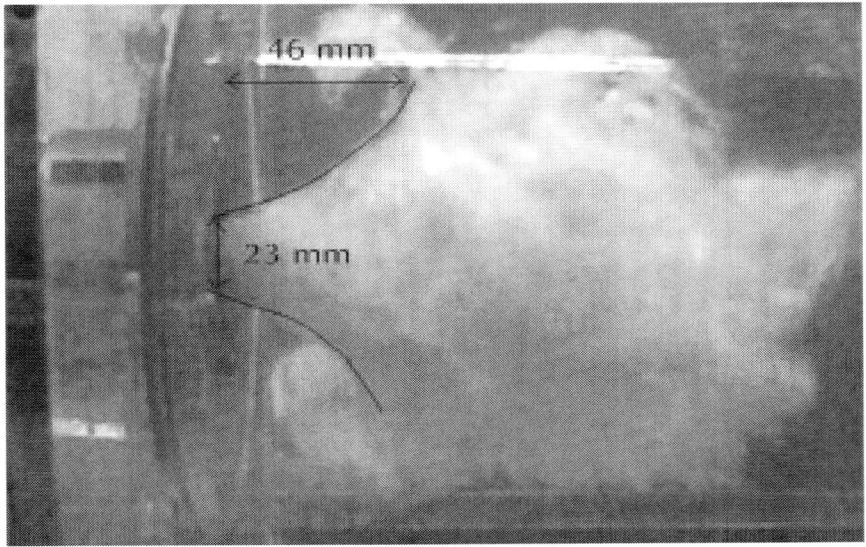

Figure 3: Photo of 60° flat guided vane swirler [4].

Type of Flame

Most of the literatures divide combustible mixture into three categories as premixed, non-premixed and partially premixed combustion. If fuel and oxidizer are mixed prior ignition, then premixed flame will propagate into the unburned reactants. If fuel and air mix at the same time and same place as they react, the diffusion or non-premixed combustion will appear. Partially premixed combustion systems are premixed flames with non-uniform fuel-oxidizer mixtures.

Gas turbines' manufacturers traditionally tend to use diffusion flame where fuel mixes with air by turbulent diffusion and the flame front stabilized in the locus of the stoichiometric mixture. The temperature of reactant is as high as 2000 °C, so the acceptable temperature at the combustor walls and turbine blades would be provide by diluted air. Although the non-premixed mixture in gas turbine combustors shows more stability in operation than premixed mixtures, but their shortcoming is high level of nitrogen oxide emission. Two most common ways of emission reduction are water injection and catalytic converter. However, the former technique is not capable of reducing NOx to the expected level at many sites, while SCR adds complexity and expense to any project.

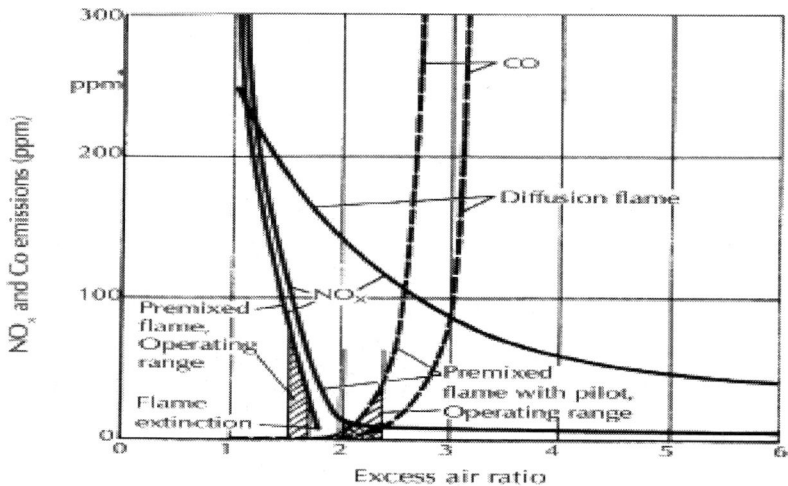

Figure 4: Operating range of premixed flames [5].

The idea of Dry Low NOx (DLN) systems proposed base on lean premixed combustion to reduce flame temperature by a non-stoichiometric mixture. Premixed systems can be operated at a much lower equivalence ratio such that the flame temperature and thermal NOx production throughout the system are decreased comparing with a diffusion system. The disadvantage of premixed systems is flame stability, especially at low equivalence ratios. Also, there is a tendency for the flame to flashback. Indeed, the current challenge of GT's developers is proposing a fuel flexible combustor for a stable combustion in all engine loads. The narrow range of fuel/air mixtures between the production of excessive NOx and excessive CO is illustrated in figure 4. NOx reduces by lowering flame temperature in a leaner mixture but CO, and unburned hydrocarbons (UHC) would increase contradictorily.

By increasing combustion residence time (volume) and preventing local quenching, CO and UHC will dissociate to CO2 and the other products. CO burns away more slowly than the other radicals, so to obtain very low level emission such as 10 ppm; it requires over 4 ms. As shown in figure 5, below 1100 °C the CO reaction becomes too slow to effectively remove the CO in an improved combustion chamber. The residence time usually does not change much on part-load because the normalized flow approximately remains constant with a variable loading.

$$NF = \dot{m}\, TP$$

$$(2)$$

Where $\dot{m}$ is the mass flow, T is combustion bulk temperature and P is combustor pressure. This will set a lower limit for the length of the primary zone in a DLN combustion system.

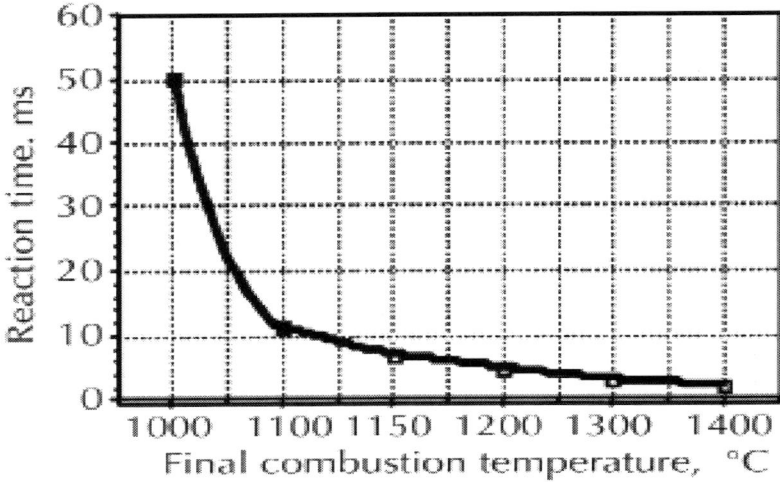

Figure 5:.Calculated reaction time to achieve a CO concentration of 10 ppm in a commercial gas turbine exhaust [6].

Fuel

One of the features of heavy-duty gas turbines is a wide fuel capability. They can operate with vast series of commercial and process by-product fuels such as natural gas, petroleum distillates, gasified coal or biomass, gas condensates, alcohols, ash-forming fuels. In a review article, Molière offered essential aspects of fuel/machine interactions in thermodynamic performance, combustion and gaseous emission [7]. To sequester and store the CO_2 of fossil fuel, some new research projects aim to assess the combustion performances of alternative fuels for clean and efficient energy production by gas turbines. Another objective is to extend the capability of dry low emission gas turbine technologies to low heat value fuels produced by gasification of biomass and H_2 enriched fuels [8-10]. Significant quantity of hydrogen in fuel has the benefit of high calorific value, but the disadvantage of high flame speed and very fast chemical times. To classify gas turbine's fuels, a common way is to split them between gas and liquid fuels, and within the gaseous fuels, to split by their calorific value as shown in table 1.

Table 1: Classification of fuels [11]

	Typical composition	Lower Heating Value kJ/Nm3	Typical specific fuels
Ultra/ Low LHV gaseous fuels	H2 < 10%	< 11,200 (< 300)	Blast furnace gas (BFG), Air blown IGCC, Biomass gasification
	CH4 < 10%		
	N2+CO > 40%		
High hydrogen gaseous fuels	H2 > 50%	5,500-11,200 (150-300)	Refinery gas, Petrochemical gas, Hydrogen power
	CxHy = 0-40%		
Medium LHV gaseous fuels	CH4 < 60%	11,200- 30,000	Weak natural gas, Landfill gas, Coke oven gas, Corex gas
	N2+CO2 = 30-50%		
	H2 = 10-50%		
Natural gas	CH4 = 90%	30,000- 45,000	Natural gas Liquefied natural gas
	CxHy = 5%		
	Inert = 5%		
High LHV gaseous fuels	CH4 and higher hydrocarbons	45,000- 190,000	Liquid petroleum gas (butane, propane) Refinery off-gas
	CxHy > 10%		
Liquid fuels	CxHy, with x > 6	32,000- 45,000	Diesel oil, Naphtha Crude oils, Residual oils, Bio-liquids

NEW COMBUSTION SYSTEMS FOR GAS TURBINES

Next-generation gas turbines will operate at higher pressure ratios and hotter turbine inlet temperatures conditions that will tend to increase nitrogen oxide emissions. To conform to future air quality requirements, lower-emitting combustion technology will be required. In this section, a number of new combustion systems have been introduced where some of them could be found in the market, and the others are under development.

Trapped Vortex Combustion (TVC)

The trapped vortex combustor (TVC) may be considered as a promising technology for both pollutant emissions and pressure drop reduction. TVC is based on mixing hot combustion products and reactants at a high rate by a cavity stabilization concept. The trapped vortex combustion concept has been under investigation since the early 1990's. The earlier studies of TVC have been concentrated on liquid fuel applications for aircraft combustors [12].

The trapped vortex technology offers several advantages as gas turbines burner:

- It is possible to burn a variety of fuels with medium and low calorific value.

- It is possible to operate at high excess air premixed regime, given the ability to support high-speed injections, which avoids flashback.

- NOx emissions reach extremely low levels without dilution or post-combustion treatments.

- Produces the extension of the flammability limits and improves flame stability.

Flame stability is achieved through the use of recirculation zones to provide a continuous ignition source which facilitates the mixing of hot combustion products with the incoming fuel and air mixture [13]. Turbulence occurring in a TVC combustion chamber is "trapped" within a cavity where reactants are injected and efficiently mixed. Since part of the combustion occurs within the recirculation zone, a "typically" flameless regime can be achieved, while a trapped turbulent vortex may provide significant pressure drop reduction [14]. Besides this, TVC is having the capability of operating as a staged combustor if the fuel is injected into both the cavities and the main airflow. Generally, staged combustion systems are having the potential of achieving about 10 to 40% reduction in NOx emissions [15]. It can also be operated as a rich-burn, quick-quench lean-burn (RQL) combustor when all of the fuel is injected into the cavities [16].

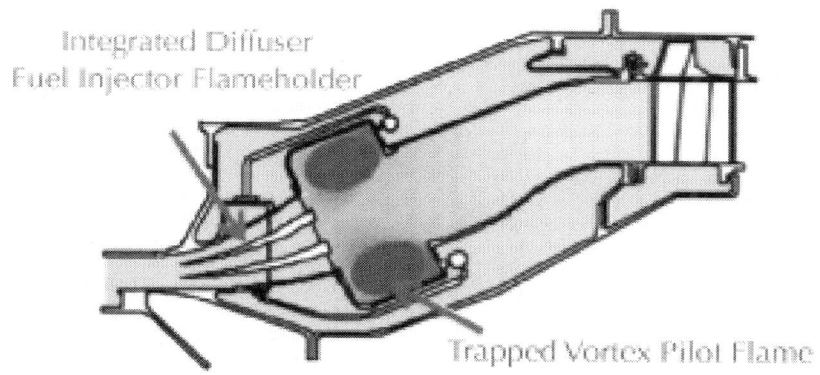

Figure 6: Trapped vortex combustor schematic.

An experiment in NASA with water injected TVC demonstrated a reduction in NOx by a factor three in a natural gas fueled and up to two in a liquid JP-8 fueled over a range in water/fuel and fuel/air ratios [17]. Replacement of natural gas fuel with syngas and hydrogen fuels has been studied numerically by Ghenai et al. [18]. The effects of secondary air jet momentum on cavity flow structure of TVC have been studied recently by Kumar and Mishra [19]. Although the actual stabilization mechanism facilitated by the TVC is relatively simple, a number of experiments and numerical simulations have been performed to enhance the stability of reacting flow inside trapped vortex. Xing et al. experimentally investigated lean blow-out of several combustors and the performance of slight temperature-raise in a single trapped vortex [20, 21]. In an experimental laboratory research, Bucher et al. proposed a new design for lean-premixed trapped vortex combustor [22].

Rich Burn, Quick- Mix, Lean Burn (RQL)

Lean direct injection (LDI) and rich-burn/quick-quench/lean-burn (RQL) are two of the prominent low-emissions concepts for gas turbines. LDI operates the primary combustion region lean, hence, adequate flame stabilization has to be ensured; RQL is rich in the primary zone with a transition to lean combustion by rapid mixing with secondary air downstream. Hence, both concepts avoid stoichiometric combustion as much as possible, but flame stabilization and combustion in the main heat release region are entirely different. Relative to aviation engines, the need for reliability and safety has led to a focus on LDI of liquid fuels [23]. However, RQL combustor technology is of growing interest for stationary gas turbines due to the attributes of more effectively processing of fuels with complex composition. The concept of RQL was proposed in 1980 as a significant effort for reducing NOx emission [24].

It is known that the primary zone of a gas turbine combustor operates most effectively with rich mixture ratios so, a "rich-burn" condition in the primary zone enhances the stability of the combustion reaction by producing and sustaining a high concentration of energetic hydrogen and hydrocarbon radical species. Secondly, rich burn conditions minimize the production of nitrogen oxides due to the relative low temperatures and low population of oxygen containing intermediate species. Critical factors of a RQL that need to be considered are careful tailoring of rich and lean equivalence ratios and very fast cooling rates. So the combustion regime shifts rapidly from rich to lean without going through the high NOx route as shown in figure 7. The drawback of this technology is increased hardware and complexity of the system.

The mixing of the injected air takes the reaction to the lean-burn zone and rapidly reduces their temperature as well. On the other hand, the temperature must be high enough to burn CO and UHC. Thus, the equivalence ratio for the lean-burn zone must be carefully selected to satisfy all emissions requirements. Typically the equivalence ratio of fuel-rich primary zone is 1.2 to 1.6 and lean-burn combustion occurs between 0.5 and 0.7 [25].

Turbulent jet in a cross-flow is an important characteristic of RQL; so many researches have been conducted to improve it. The mixing limitation in a design of RQL/TVC combustion system addressed by Straub et al. [26]. Coaxial swirling air discussed experimentally by Cozzi and Coghe [27]. Furthermore, an experimental study of the effects of elevated pressure and temperature on jet mixing and emissions in an RQL reported by Jermakian et al. [28]. Fuel flexible combustion with RQL system is an interest of turbine manufacturer. GE reported results of a RQL test stand in their integrated gasification combined cycle (IGCC) power plants program [29, 30]. The test of Siemens-Westinghouse Multi-Annular Swirl Burner (MASB) was successfully performed at the University of Tennessee Space Institute in Tullahoma [31]. Others, such as references [32-35] utilize CFD to investigate the performance of RQL combustor.

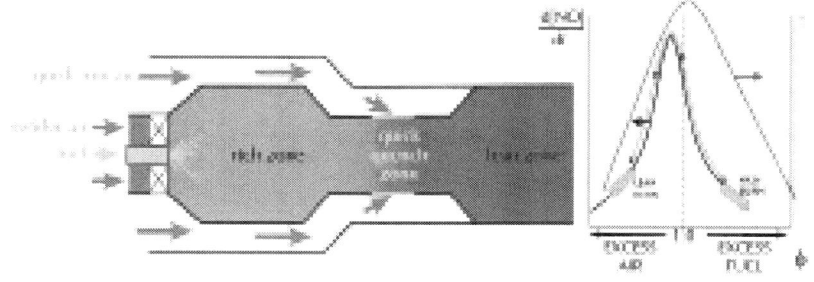

Figure 7: Rich-Burn, Quick-Mix, Lean-Burn combustor.

Staged Air Combustion

The COSTAIR[4] - combustion concept uses continuously staged air and internal recirculation within the combustion chamber to obtain a stable combustion with low NOX and CO emissions. Research work on staged combustors started in the early 1970s under of the Energy Efficient Engine (E^3) Program in the USA [36] and now widely used in industrial engines burning gaseous fuels, in both axial and radial configurations. The aero-derived GE LM6000 and CFM56-5B as well as RR211 DLE industrial engine employ staged combustion of premixed gaseous fuel/air mixtures. Recently, a research project proposed a COSTAIR burner system optimized for low calorific gases within a micro gas turbine [37].

The principle of staged air combustion is illustrated in Figure 8. It consists of a coaxial tube; the combustion air flows through the inner tube and the fuel through the outer cylinder ring. The combustion air is continually distributed throughout the combustion chamber by an air distributor with numerous openings on its contour, and fuel enters by several jets arranged around the air distributor.

The COSTAIR burner has the advantages of operating in full diffusion mode or in partially premixed mode. The heat is released more uniformly throughout the combustion chamber also the recirculated gas absorb some of the heat of combustion. It capable to work stable at cold combustor walls as well as high air ratio. Experimental measurements show that this combustion system allows clean exhaust. For instance, in an experimental research project of European Commission [39], NOx emission values was in the range of 2-4 ppm at an air ratio of 2.5 over different loading. Furthermore, the corresponding CO emission was less than 7 ppm.

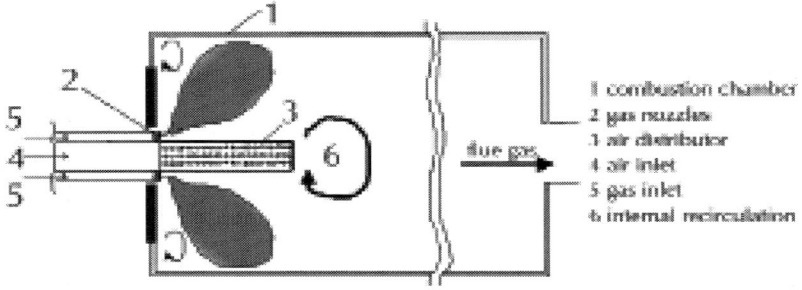

Figure 8: COSTAIR combustion concept [38].

Staged combustion can occur in either a radial or axial pattern, but in either case the goal is to design each stage to optimize particular performance aspects. The main advantages or major drawbacks of each type have been discussed by Lefebvre [25].

Mild Combustion

Heat recirculating combustion was clearly described by Weinberg as a concept for improving the thermal efficiency [40]. In 1989, a surprising phenomenon was observed during experiments with a self-recuperative burner. At furnace temperatures of 1000°C and about 650°C air preheated temperature; no flame could be seen, but the fuel was completely burnt. Furthermore, the CO and NOx emissions from the furnace were considerably low [41]. Different combustion zones against rate of dilution and oxygen content is shown infigure 9. In flameless combustion, the oxidation of fuel occurs with a very limited oxygen supply at a very high temperature. Spontaneous ignition occurs and progresses with no visible or audible signs of the flames usually associated with burning. The chemical reaction zone is quite diffuse, and this leads to almost uniform heat release and a smooth temperature profile. All these factors could result in a much more efficient process as well as reducing emissions.

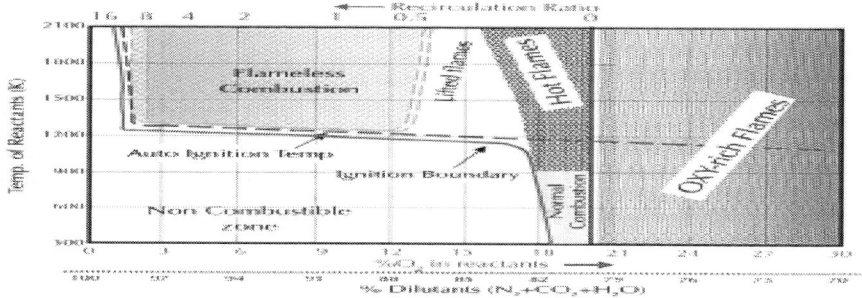

Figure 9: Different combustion regimes [64].

Flameless combustion is defined where the reactants exceed self-ignition temperature as well as entrain enough inert combustion products to reduce the final reaction temperature [42]. In the other word, the essence of this technology is that fuel is oxidized in an environment that contains a substantial amount of inert (flue) gases and some, typically not more than 3–5%, oxygen. Several different expressions are used to identify similar though such as HiTAC[5] - , HiCOT[6] - , MILD[7] - combustion, FLOX[8] - and CDC[9] - . HiTAC refers to increase the air temperature by preheating systems such as regenerators. HiCOT commonly belongs to the wider sense, which exploits high-temperature reactants; therefore, it is not limited to air. A combustion process is named FLOX or MILD when the inlet temperature of the main reactant flow is higher than mixture autoignition temperature and the maximum allowable temperature increase during combustion is lower than mixture autoignition temperature, due to dilution [42]. The common key feature to achieve reactions in CDC mode (non-premixed conditions) is the separation and controlled mixing of higher momentum air jet and the lower momentum fuel jet, large amount of gas recirculation and higher turbulent mixing rates to achieve spontaneous ignition of the fuel to provide distributed combustion reactions [43].Figure 10 schematically shows a comparison between conventional burner and flameless combustion.

To recap, the main characteristics of flameless oxidation combustion are:

- Recirculation of combustion products at high temperature (normally > 1000 °C),
- Reduced oxygen concentration at the reactance,
- Low Damköhler number (Da[10] -),
- Low stable adiabatic flame temperature,
- Reduce temperature peaks,
- Highly transparent flame,
- Low acoustic oscillation and
- Low NOx and CO emissions.

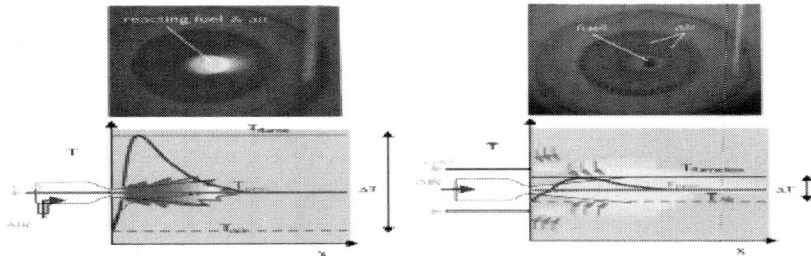

Figure 10: Flame (left) and flameless (right) firing.

In spite of a number of activities for industrial furnaces, the application of flameless combustion in the gas-turbine combustion system is in the preliminary phase [44]. The results from techno-economic analysis of Wang et al. showed that the COSTAIR and FLOX cases had technical and economic advantages over SCR [45]. Luckerath, R., et al., investigated flameless combustion in forward flow configuration in elevated pressure up to 20atm for application to gas turbine combustors [44, 46]. In a novel design of Levy et al. that named FLOXCOM, flameless concept has been proposed for gas turbines by establishing large recirculation zone in the combustion chamber [47, 48]. Lammel et al. developed a FLOX combustion at high power density and achieved low NOx and CO levels [49]. The concept of colorless distributed combustion has been demonstrated by Gupta et al. for gas turbine application in a number of publications [43, 50-55].

Surface Stabilized Combustion

One specification of gas turbine combustor is higher thermal intensity range (at least 5 MW/m^3-atm) than industrial furnaces which operate at thermal intensity of less than 1 MW/m^3-atm. Therefore, designs of gas turbine's combustors are based on turbulent flow concept, except a technology named NanoSTAR from Alzeta Corporation. Alzeta reported the proof-of-concept of high thermal intensity laminar surface stabilized flame by using a porous metal-fiber mat since 2001 [56-58]. Lean premixed combustion technology is limited by the apparition of combustion instabilities, which induce high pressure fluctuations, which can produce turbine damage, flame extinction, and CO emissions [59]. However, full scale test of NanoSTAR demonstrated low emissions performance, robust ignition and extended turndown ratio [60]. In particular, the following characteristics form the key specifications of NanoSTAR for distributed power generation gas turbine combustors [61]:

- The combustor fuel is limited to natural gas.
- Total combustor pressure drop limited to 2-4% of the system pressure.
- Operation at combustion air preheat temperatures up to 1150°F.
- Volumetric firing rates approaching 2 MMBtu/hr/atm/ft^3.
- Turbine Rotor Inlet Temperatures (TRIT) over 2200°F (valid for the Mercury 50, although Allison has operated combustors at 2600°F).
- Operation with axial combustors or external can combustors. Expected component lifetimes of 30,000 hours for industrial turbines.

A single prototype burner Porous burner which sized to fit inside an annular combustion liner (about 2.5 inches in diameter by 7 inches in length) is shown in figure 11 with its arrangement in a typical combustor.

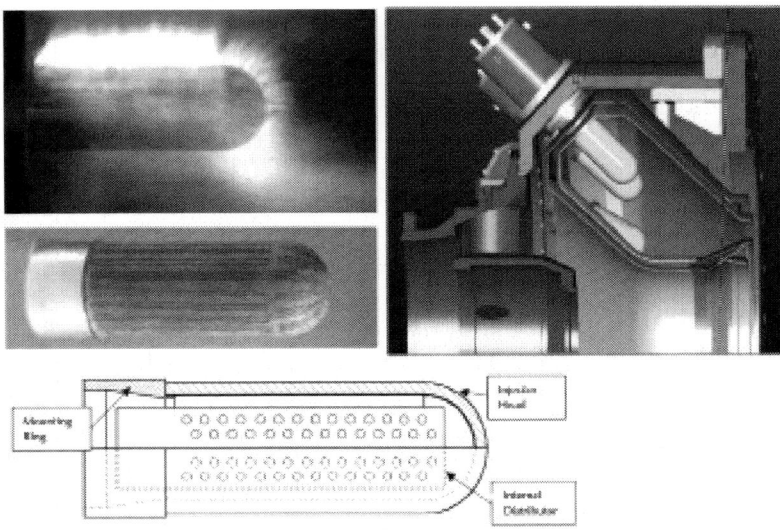

Figure 11: NanoSTAR burner and its arrangement in a canted combustion system [62].

The operation of this type of surface stabilized combustion is characterized by the schematic in Figure 12(left), which shows premixed fuel and air passing through the metal fiber mat in two distinct zones. Premixed fuel comes through the low conductivity porous and burns in narrow zones, A, as it leaves the surface. Under lean conditions this will manifest as very short laminar flamelets, but under rich conditions the surface combustion will become a diffusion dominated reaction stabilized just over a millimeter above the metal matrix, which proceeds without visible flame and heats the outer surface of the mat to incandescence. Secondly, adjacent to these radiant zones, the porous plate is perforated to allow a high flow of the premixed fuel and air. This flow forms a high intensity flame, B, stabilized by the radiant zones so, it is possible to achieve very high fluxes of energy, up to 2MMBtu/hr/ft^2 [63]. A picture of an atmospheric burner in operation clearly shows the technology in action (right of figure 12).

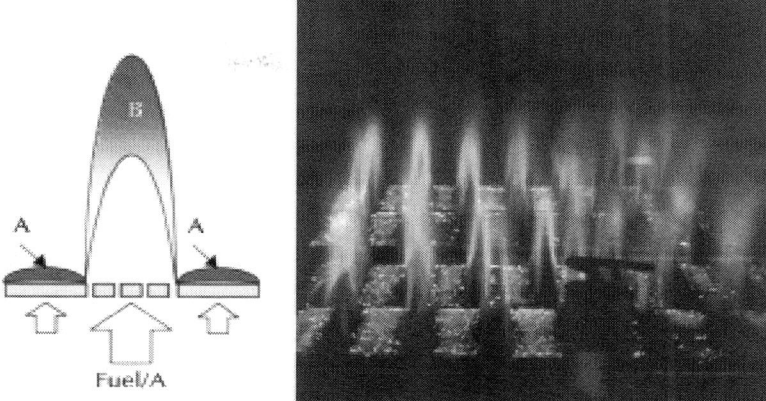

Figure 12: Surface stabilized burner pad firing at atmospheric conditions.

The specific perforation arrangement and pattern control the size and shape of the laminar flamelets. The perforated zones operate at flow velocities of up to 10 times the laminar flame speed producing a factor of ten stretch of the flame surface and resulting in a large laminar flamelets. The alternating arrangement of laminar blue flames and surface combustion, allows high firing rates to be achieved before flame liftoff occurs, with the surface combustion stabilizing the long laminar flames by providing a pool of hot combustion radicals at the flame edges.

CONCLUSIONS

A review of technologies for reducing NOx emissions as well as increasing thermal efficiency and improving combustion stability has been reported here. Trade-offs when installing low NOx burners in gas turbines include the potential for decreased flame stability, reduced operating range and more strict fuel quality specifications. In the other word, although, the turbine inlet temperature is the major factor determining the overall efficiency of the gas turbine but higher inlet temperatures will result in larger NOx emissions. So the essential requirement of new combustor design is a trade-off between low NOx and improved efficiency.

REFERENCES

1. Chambers A and Trottier S (2007) Technologies for Reducing Nox Emissions from Gas-Fired Stationary Combustion Sources. Alberta Research Council, Edmonton, Canada

2. Kiel B, Garwick LK, Gord JR, Miller J, Lynch A, Hill R and Phillips S (2007) A Detailed Investigation of Bluff Body Stabilized Flames. 45th AIAA Aerospace Sciences Meeting and Exhibit.

3. Gupta AK, Lilley DG and Syred N (1984) Swirl Flows, Abacus Press.

4. Jaafar MNM, Jusoff K, Osman MS and Ishak MSA (2011) Combustor Aerodynamic Using Radial Swirler. International Journal of the Physical Sciences. 6: 3091 - 3098.

5. Moore MJ (1997) Nox Emission Control in Gas Turbines for Combined Cycle Gas Turbine Plant. Proceedings of the Institution of Mechanical Engineers, Part A: Journal of Power and Energy. 211: 43-52.

6. Kajita S and Dalla Betta R (2003) Achieving Ultra Low Emissions in a Commercial 1.4 Mw Gas Turbine Utilizing Catalytic Combustion. Catalysis Today. 83: 279-288.

7. Molière M (2000) Stationary Gas Turbines and Primary Energies: A Review of Fuel Influence on Energy and Combustion Performances. International Journal of Thermal Sciences. 39: 141-172.

8. Gupta KK, Rehman A and Sarviya RM (2010) Bio-Fuels for the Gas Turbine: A Review. Renewable and Sustainable Energy Reviews. 14: 2946-2955.

9. Gökalp I and Lebas E (2004) Alternative Fuels for Industrial Gas Turbines (Aftur). Applied Thermal Engineering. 24: 1655-1663.

10. Juste GL (2006) Hydrogen Injection as Additional Fuel in Gas Turbine Combustor. Evaluation of Effects. International Journal of Hydrogen Energy. 31: 2112-2121.

11. Jones R, Goldmeer J and Monetti B (2011) Addressing Gas Turbine Fuel Flexibility. GE Energy

12. Haynes J, Janssen J, Russell C and Huffman M (2006) Advanced Combustion Systems for Next Generation Gas Turbines. United States. Dept. of Energy, Washington, D.C.; Oak Ridge, Tenn.

13. Sturgess GJ and Hsu KY (1998) Combustion Characteristics of a Trapped Vortex Combustor. Applied vehicle technology panel symposium.

14. Bruno C and Losurdo M (2007) The Trapped Vortex Combustor: An Advanced Combustion Technology for Aerospace and Gas Turbine Applications. In: Syred N and Khalatov A, Syred N and Khalatov A, editors. Advanced Combustion and Aerothermal Technologies. Springer Netherlands, pp 365-384.

15. Mishra DP (2008) Fundamentals of Combustion, Prentice-Hall Of India Pvt. Limited.

16. Acharya S, Mancilla PC and Chakka P (2001) Performance of a Trapped Vortex Spray Combustor. ASME International Gas Turbine Conference. ASME.

17. Hendricks RC, Shouse DT and Roquemore WM (2005) Water Injected Turbomachinery. NASA, Glenn Research Center

18. Ghenai C, Zbeeb K and Janajreh I (2012) Combustion of Alternative Fuels in Vortex Trapped Combustor. Energy Conversion and Management. In press

19. Ezhil Kumar PK and Mishra DP (2011) Numerical Simulation of Cavity Flow Structure in an Axisymmetric Trapped Vortex Combustor. Aerospace Science and Technology. In Press

20. Xing F, Wang P, Zhang S, Zou J, Zheng Y, Zhang R and Fan W (2012) Experiment and Simulation Study on Lean Blow-out of Trapped Vortex Combustor with Various Aspect Ratios. Aerospace Science and Technology. 18: 48-55.

21. Xing F, Zhang S, Wang P and Fan W (2010) Experimental Investigation of a Single Trapped-Vortex Combustor with a Slight Temperature Raise. Aerospace Science and Technology. 14: 520-525.

22. Bucher J, Edmonds RG, Steele RC, Kendrick DW, Chenevert BC and Malte PC (2003) The Development of a Lean-Premixed Trapped Vortex Combustor. ASME Turbo Expo 2003 Power for Land, Sea, and Air.

23. Dunn-Rankin D (2008) Lean Combustion: Technology and Control, Academic Press, USA.

24. Mosier SA and Pierce RM (1980) Advanced Combustion Systems for Stationary Gas Turbine Engines. Volume I. Review and

Preliminary Evaluation. Final Report December 1975-September 1976. pp Medium: X; Size: Pages: 49.

25. Lefebvre AH and Ballal DR (2010) Gas Turbine Combustion: Alternative Fuels and Emissions, Taylor & Francis.

26. Straub DL, Casleton KH, Lewis RE, Sidwell TG, Maloney DJ and Richards GA (2005) Assessment of Rich-Burn, Quick-Mix, Lean-Burn Trapped Vortex Combustor for Stationary Gas Turbines. Journal of engineering for gas turbines and power. 127: 36-41.

27. Cozzi F and Coghe A (2012) Effect of Air Staging on a Coaxial Swirled Natural Gas Flame. Experimental Thermal and Fluid Science. In press

28. Jermakian V, McDonell VG and Samuelsen GS (2012) Experimental Study of the Effects of Elevated Pressure and Temperature on Jet Mixing and Emissions in an Rql Combustor for Stable, Efficient and Low Emissions Gas Turbine Applications. Advanced Power and Energy Program, University of California, Irvine

29. Feitelberg AS, Jackson MR, Lacey MA, Manning KS and Ritter AM (1996) Design and Performance of a Low Btu Fuel Rich-Quench-Lean Gas Turbine Combustor. Advanced coal-fired power systems review meeting. USA DOE, Morgantown Energy Technology Center.

30. Feitelberg AS and Lacey MA (1998) The Ge Rich-Quench-Lean Gas Turbine Combustor. Journal of Engineering for Gas Turbines and Power, Transactions of the ASME. 120: 502-508.

31. Brushwood J (1999) Syngas Combustor for Fluidized Bed Applications 15th Annual Fluidized Bed Conference.

32. Howe GW, Li Z, Shih TI-P and Nguyen HL (1991) Simulation of Mixing in the Quick Quench Region of a Richburn-Quick Quench Mix-Lean Burn Combustor. 29th Aerospace Sci Meeting. AIAA.

33. Cline MC, Micklow GJ, Yang SL and Nguyen HL (1992) Numerical Analysis of the Flow Fields in a Rql Gas Turbine Combustor.

34. Talpallikar MV, Smith CE, Lai MC and Holdeman JD (1992) Cfd Analysis of Jet Mixing in Low Nox Flametube Combustors. Journal of Engineering for Gas Turbines and Power. 114: 416-424.

35. Blomeyer M, Krautkremer B, Hennecke DK and Doerr T (1999) Mixing Zone Optimization of a Rich-Burn/Quick-Mix/Lean-Burn Combustor. Journal of Propulsion and Power 15: 288-303.

36. Wulff A and Hourmouziadis J (1997) Technology Review of Aeroengine Pollutant Emissions. Aerospace Science and Technology. 1: 557-572.

37. Leicher J, Giese A, Görner K, Scherer V and Schulzke T (2011) Developing a Burner System for Low Calorific Gases in Micro Gas Turbines: An Application for Small Scale Decentralized Heat and Power Generation International Gas Union Research Conference.

38. Al-Halbouni A, Flamme M, Giese A, Scherer V, Michalski B and Wünning JG (2004) New Burner Systems with High Fuel Flexibility for Gas Turbines. 2nd International Conference on Industrial Gas Turbine Technologies.

39. Flamme M (2004) New Combustion Systems for Gas Turbines (Ngt). Applied Thermal Engineering. 24: 1551-1559.

40. WEINBERG F (1996) Heat-Recirculating Burners : Principles and Some Recent Developments. Combustion Science and Technology. 121: 3-22.

41. Wünning J (2005) Flameless Oxidation. 6th HiTACG Symposium.

42. Cavaliere A and de Joannon M (2004) Mild Combustion. Progress in Energy and Combustion Science. 30: 329-366.

43. Arghode VK, Gupta AK and Bryden KM (2012) High Intensity Colorless Distributed Combustion for Ultra Low Emissions and Enhanced Performance. Applied Energy. 92: 822-830.

44. Li P, Mi J, Dally B, Wang F, Wang L, Liu Z, Chen S and Zheng C (2011) Progress and Recent Trend in Mild Combustion. SCIENCE CHINA Technological Sciences. 54: 255-269.

45. Wang YD, Huang Y, McIlveen-Wright D, McMullan J, Hewitt N, Eames P and Rezvani S (2006) A Techno-Economic Analysis of the Application of Continuous Staged-Combustion and Flameless Oxidation to the Combustor Design in Gas Turbines. Fuel Processing Technology. 87: 727-736.

46. Luckerath R, Meier W and Aigner M (2008) Flox Combustion at High Pressure with Different Fuel Compositions. Journal of Engineering for Gas Turbines and Power. 130: 011505.

47. Costa M, Melo M, Sousa J and Levy Y (2009) Experimental Investigation of a Novel Combustor Model for Gas Turbines. Journal of Propulsion and Power 25: 609-617.

48. Levy Y, Sherbaum V and Arfi P (2004) Basic Thermodynamics of Floxcom, the Low-Nox Gas Turbines Adiabatic Combustor. Applied Thermal Engineering. 24: 1593-1605.

49. Lammel O, Schutz H, Schmitz G, Luckerath R, Stohr M, Noll B, Aigner M, Hase M and Krebs W (2010) Flox Combustion at High Power Density and High Flame Temperatures. Journal of Engineering for Gas Turbines and Power. 132: 121503.

50. Arghode VK and Gupta AK (2011) Development of High Intensity Cdc Combustor for Gas Turbine Engines. Applied Energy. 88: 963-973.

51. Khalil AEE and Gupta AK (2011) Distributed Swirl Combustion for Gas Turbine Application. Applied Energy. 88: 4898-4907.

52. Arghode VK and Gupta AK (2010) Effect of Flow Field for Colorless Distributed Combustion (Cdc) for Gas Turbine Combustion. Applied Energy. 87: 1631-1640.

53. Arghode VK, Khalil AEE and Gupta AK (2012) Fuel Dilution and Liquid Fuel Operational Effects on Ultra-High Thermal Intensity Distributed Combustor. Applied Energy. 95: 132-138.

54. Khalil AEE, Arghode VK, Gupta AK and Lee SC (2012) Low Calorific Value Fuelled Distributed Combustion with Swirl for Gas Turbine Applications. Applied Energy. 98: 69-78.

55. Khalil AEE and Gupta AK (2011) Swirling Distributed Combustion for Clean Energy Conversion in Gas Turbine Applications. Applied Energy. 88: 3685-3693.

56. Greenberg SJ, McDougald NK and Arellano LO (2004) Full-Scale Demonstration of Surface-Stabilized Fuel Injectors for Sub-Three Ppm Nox Emissions. ASME Conference Proceedings. 2004: 393-401.

57. Greenberg SJ, McDougald NK, Weakley CK, Kendall RM and Arellano LO (2003) Surface-Stabilized Fuel Injectors with Sub-Three Ppm Nox Emissions for a 5.5 Mw Gas Turbine Engine. International Gas Turbine and Aeroengine Congress and Exhibition. American Society of Mechanical Engineers.

58. Weakley CK, Greenberg SJ, Kendall RM, McDougald NK and Arellano LO (2002) Development of Surface-Stabilized Fuel Injectors with Sub-Three Ppm Nox Emissions. International Joint Power Generation Conference. American Society of Mechanical Engineers.

59. Cabot G, Vauchelles D, Taupin B and Boukhalfa A (2004) Experimental Study of Lean Premixed Turbulent Combustion in a Scale Gas Turbine Chamber. Experimental Thermal and Fluid Science. 28: 683-690.

60. Mcdougald NK (2005) Development and Demonstration of an Ultra Low Nox Combustor for Gas Turbines. USA DOE, Office of Energy Efficiency and Renewable Energy, Washington, D.C; Oak Ridge, Tenn.

61. Arellano LO, Bhattacharya AK, Smith KO, Greenberg SJ and McDougald NK (2006) Development and Demonstration of Engine-Ready Surface-Stabilized Combustion System. ASME Turbo Expo 2006: Power for Land, Sea, and Air.

62. Arellano L, Smith KO, California Energy Commission. Public Interest Energy R and Solar Turbines I (2008) Catalytic Combustor-Fired Industrial Gas Turbine Pier Final Project Report, California Energy Commission, [Sacramento, Calif.].

63. Clark H, Sullivan JD, California Energy Commission. Public Interest Energy R, California Energy Commission. Energy Innovations Small Grant P and Alzeta C (2001) Improved Operational Turndown of an Ultra-Low Emission Gas Turbine Combustor, California Energy Commission, Sacramento, Calif.

64. Arvind G. Rao and Yeshayahou Levy, "A New Combustion Methodology for Low Emission Gas Turbine Engines", 8th HiTACG conference, July 5-8 2010, Poznan.

Citations

CHAPTER 1

John Oakey; Nigel Simms; Paul Kilgallon, Gas Turbines: Gas Cleaning Requirements for Biomass-Fired Systems, doi.org/10.1590/S1516-14392004000100004.

CHAPTER 2

Melissa Wilcox, Rainer Kurz, and Klaus Brun, "Technology Review of Modern Gas Turbine Inlet Filtration Systems,"International Journal of Rotating Machinery, vol. 2012, Article ID 128134, 15 pages, 2012. doi:10.1155/2012/128134.

CHAPTER 3

Valceres V. R. e Silva, Wael Khatib, and Peter J. Fleming, Control system design for a gas turbine engine using evolutionary computing for multi-disciplinary optimization, doi: 10.1590/S0103-17592007000400007.

CHAPTER 4

Ulrich Kueppers, Corrado Cimarelli, Kai-Uwe Hess, Jacopo Taddeucci, Fabian B Wadsworth, and Donald B Dingwell, The thermal stability of Eyjafjallajökull ash versus turbine ingestion test sands, doi:10.1186/2191-5040-3-4.

CHAPTER 5

Maher M Abou Al-Sood, Kassem K Matrawy, and Yousef M Abdel-Rahim, Optimum Parametric Performance Characterization of an Irreversible Gas Turbine Brayton Cycle, doi:10.1186/2251-6832-4-37.

CHAPTER 6

Takeharu Hasegawa (2013). Development of Semiclosed Cycle Gas Turbine for Oxy-Fuel IGCC Power Generation with CO_2 Capture, Progress in Gas Turbine Performance, Dr. Ernesto Benini (Ed.), ISBN: 978-953-51-1166-5, InTech, DOI: 10.5772/54406.

CHAPTER 7

Ene Barbu, Romulus Petcu, Valeriu Vilag, Valentin Silivestru, Tudor Prisecaru, Jeni Popescu, Cleopatra Cuciumita and Sorin Tomescu (2013). Gas Turbine Cogeneration Groups Flexibility to Classical and Alternative Gaseous Fuels Combustion, Progress in Gas Turbine Performance, Dr. Ernesto Benini (Ed.), ISBN: 978-953-51-1166-5, InTech, DOI: 10.5772/54404.

CHAPTER 8

M. Khosravy el_Hossaini (2013). Review of the New Combustion Technologies in Modern Gas Turbines, Progress in Gas Turbine Performance, Dr. Ernesto Benini (Ed.), ISBN: 978-953-51-1166-5, InTech, DOI: 10.5772/54403.

Index

A

Arizona Test Dust (ATD) 89

B

Back work ratio (BWR) 114

C

Central Research Institute of Electric Power Industry (CRIEPI) 138, 165

Collaborative optimization (CO) 70

Combustion chamber 207, 209, 210, 213, 216, 219, 220, 222

D

Differential scanning calorimetry (DSC) 91

Dry Low NOx (DLN) 213

E

Ecological coefficient of performance (ECOP) 114

Engine pressure ratio (EPR) 66

Erosion 25

Evolutionary algorithms (EAs) 64

Evolutionary programming (EP) 64

Evolution strategies (ES) 64

F

Filter 27, 28, 30, 33, 41, 43, 60
Filter efficiency 29, 30
First gas turbine 108
Fouling 25, 59, 60

G

Gas turbine 2, 5, 12, 13, 14, 24,
    45, 49, 107, 108, 109, 126,
    130, 169, 170, 171, 172,
    173, 174, 175, 176, 177,
    178, 180, 183, 184, 185,
    186, 187, 188, 189, 190,
    192, 195, 196, 197, 198,
    199, 200, 201, 202, 203,
    204, 205, 206
Gas turbine engines (GTE) 65
General mathematical 107, 130
Genetic programming (GP) 64
Glass transition 97, 98, 99
Global warming 1, 2

H

High pressure 4, 18
High temperature 112, 116, 122

I

Ingestion of airborne volcanic
    matter 81
Inlet guide vane angle (IGV) 65,
    66
Inlet Guide Vanes (IGVs) 39
Integrated coal gasification com-
    bined cycle (IGCC) 138
Integrated gasification combined
    cycle (IGCC) 219
International civil aviation orga-
    nization [ICAO] 83, 103

L

Lean direct injection (LDI) 217
Life Cycle Cost (LCC) 55
Liquefied petroleum gas (LPG)
    177
Lower heating value (LHV) 175,
    178, 181

M

MIMO (Multi Input Multi Output)
    70
Minimum Efficiency Reporting
    Value (MERV) 30
Mitsubishi Heavy Industries Ltd.
    (MHI) 172
MOGA (multiobjective genetic
    algorithm) 71
Monte Carlo methodology
    (MCM) 115
Most Penetrating Particle Size ef-
    ficiency (MPPS) 31
Multidisciplinary optimization
    (MDO) 63
Multiobjective genetic algorithm
    (MOGA) 68
Multiobjective optimization (MO)
    65, 68

N

Natural gas 13
Natural gas (NG) 137
Net Present Value (NPV) 56

P

PI (proportional and integral) 65,
    66
Power plant 108

R

Rich-burn/quick-quench/lean-
burn (RQL) 217

T

Thermogravimetric (TG) 91
trapped vortex combustor (TVC)
216
Turbine Rotor Inlet Temperatures
(TRIT) 223

U

Unburned hydrocarbons (UHC)
213

W

Wide range 5

X

X-Ray Fluorescence (XRF) 87